Anonymous

Hints About Heating

Suggestions respecting hot-air furnace work, together with tables of

dimensions, capacities, etc.

Anonymous

Hints About Heating
Suggestions respecting hot-air furnace work, together with tables of dimensions, capacities, etc.

ISBN/EAN: 9783337303938

Printed in Europe, USA, Canada, Australia, Japan

Cover: Foto ©berggeist007 / pixelio.de

More available books at **www.hansebooks.com**

HINTS ⚜ ABOUT ⚜ HEATING

Containing valuable suggestions respecting hot-air furnace work, together with tables of dimensions, capacities, etc., prepared with especial reference to the Paragon Hot Air Furnace

Part One ❧ Third Edition ❧ Revised and Enlarged

Philadelphia and Baltimore

ISAAC A. SHEPPARD & CO

Mdcccxcviii

HINTS ABOUT HEATING

THIS pamphlet is not intended as a manual of information upon the subject of heating by hot air, but simply to point out to purchasers some of the requisites of satisfactory work, and to assist any dealer, who may be without experience in furnace work, to give satisfaction to his customers.

Our climate requires more or less artificial heat during the greater part of the year. If the family is to keep in good health, proper warmth and ventilation in the dwelling are essential; and money spent in securing comfort in these respects will often prevent sickness and save doctors' bills.

Open Fireplaces Insufficient

Open fireplaces are no longer regarded, in this country, as anything more than a pleasant means of supplementing heat derived from other sources. When in actual use, they afford excellent ventilation. For this reason, it is well to provide them in every dwelling. For heating purposes, however, the main reliance must be placed upon either steam or hot water apparatus, or upon Hot Air Furnaces.

Advantages of Warm Air Heating

Steam heating, whether direct or indirect, although well adapted to the requirements of large public institutions, is less suited to ordinary buildings. It is costly ; it requires skill and good judgment in its management; and it calls for constant attention. Hot water apparatus, while more safe and more easily managed than steam, is more expensive. Indirect steam or hot water heating, moreover, is exceedingly costly; and the use of direct radiation relies upon heating, over and over again, the air that is *already in the room*. It must certainly be conceded, that a more healthful method of house warming is one which furnishes a constant supply of fresh, pure air, *taken from the outside atmosphere* and thoroughly warmed before entering the room. This is exactly what is accomplished by a well constructed Hot Air Furnace. The combination of heating by hot air and by hot water also, from the same source of heat, obviates the disadvantages which attend heating by direct radiation alone. This mode of heating will be treated of later.

Healthfulness of Warm Air Heating

Upon the grounds of superior healthfulness, safety, economy in first cost, ease of management and inexpensiveness of repairs, a good Hot Air Furnace is to be preferred to all other forms of heating apparatus, whenever its use is feasible. No objections have ever been urged against Hot Air Furnaces that cannot easily be shown to originate either in defective construction or in improper management. For the latter, no furnace can justly be blamed. As to the former, it can only be said that furnaces that are defective in construction can always be obtained by those who are unwilling to pay for a good one. On the other hand, it is also true that a good furnace, satisfactorily put up, is within the reach of every person who is willing to pay a fair price. It is to the interest of both the furnace manufacturer and the furnace seller to do their best to satisfy a purchaser who is willing to compensate them reasonably for their outlay.

The Best is the Cheapest

This is a time worn proverb; but it is emphatically true when applied to Hot Air Furnaces.   It is an unreasoning and false economy that leads house owners to use a " cheap" type of furnace, put up in a "cheap" way.   Good work, in any branch of manufacture, cannot be obtained without paying for it what it is worth. Surely, the health and comfort of one's family are matters of great importance; and those persons who are planning to heat their own homes will not find it to their interest, in the long run, to use poor furnaces improperly set.   If they will not pay the furnace man, they may have to pay the doctor, and the coal dealer also.

Even in the case of the houses so often built in our great cities, in long rows, upon speculation, with the intention of selling as quickly as possible, it is to the interest of builders to get good work in this line. Good furnace work will enhance the value of the property, and will help it to an earlier sale, at a better price than if this important essential were slighted.

It is assumed that those who read these pages want good work and are willing to pay a reasonable price for it.

Furnace Problems of Two Kinds

The problems that arise in furnace work are of two kinds, namely:—those that relate to the *production* of heat, and those that relate to its *proper distribution*.

The furnace used is responsible *only for the solution of the former*, and even then only when properly managed. The solution of all the problems that relate to the *proper distribution* of the heat supplied by the furnace *rests with the person who sets the furnace*. He decides upon its location, adjusts the hot air pipes and flues, determines upon their sizes, locates the registers and provides for cold air supply. He needs to have not a little good judgment, experience and mechanical skill; for the successful heating of a building depends quite as much upon proper attention to each of these matters as upon the heating capacity of the furnace. Nothing is more common than to find a furnace complained of, when the trouble is entirely due to defects in the mode of distributing the heat produced by it, the arrangements made for this purpose being so insufficient as to make it an impossibility for the hot air generated by the furnace to pass from the furnace to the rooms in which the heat is desired.

Distribution of Heat First Considered

The principles that govern the proper distribution of heated air are few; but their application differs more or less in each specific case. Much experience and ingenuity are at times necessary in order to attain the best results. We shall defer, for the present, the discussion of such matters as relate to the production of heat, and shall first consider the mode of effecting a *proper distribution* of the warm air generated by a furnace.

Movement of Heated Air

Three fundamental facts must be remembered:

I. Heated air is set in motion *by the pressure of cold air beneath it.*

II. Heated air always moves most readily *in the direction in which it meets the least resistance.*

III. The velocity of heated air in a flue *increases* in proportion to the height of the flue and its excess of temperature over that of the outside air.

Upon the observance of these facts all satisfactory hot air heating depends. From the first, we learn the need of a *proper cold air supply.* When the other two are borne in mind, it is apparent that warm air will move more easily in a *vertical* than in a *horizontal* direction, through *short* horizontal pipes more easily than through *long* ones, through *large* pipes more easily than through *small* ones, through *round or square* pipes more easily than through *flat* ones, and more easily through *curved* than through *right angled* elbows. Also, it appears that warm air will move *with the prevailing wind* rather than against it, into a *well ventilated* room rather than into a *close* one, and into an *upper* room in preference to a *lower* one.

The bearing of these well established facts upon the work of intelligent and satisfactory furnace setting, will be seen as the discussion of the subject proceeds.

Location of Furnace

The furnace should always be placed where it will be as easy as possible for the warm air to pass *quickly* and *uniformly* to the rooms that are to be heated by it. Generally speaking, a central position is the most favorable for this purpose; as it causes the lines of pipe to the different hot air flues and registers to be as nearly as possible of *equal length.* This makes the *elevation* of the several pipes as nearly *equal* as possible. Other things being equal, uniformity in distribution is thereby secured.   The greater the elevation of a pipe the more easily will the hot air pass through it, and the *shorter* the pipe the *greater its elevation;* so that if a furnace be so placed that some of the pipes are very short and others very long, the short pipes will tend to carry away most of the heat and the long ones will get very little. In cases in which this arrangement cannot be avoided, the short pipes should be made *smaller in size* than the long ones, in order to counteract this tendency.

Heated air always moves slowly and with difficulty through pipes that are horizontal, or nearly so; and hot air pipes should never have an elevation less than 1½ inches per running foot.   If the cellar is too low to give such elevation to the pipes, the furnace must be placed in a pit of sufficient depth, lined with brick laid in cement.   If the cellar should be damp, the pit should be drained into a drainage well of a greater depth.

That a furnace should be centrally located is not an invariable rule; but it is to be advised in the case of such buildings as are well sheltered from the winter winds.   When the exposure of a building is great, as in the case of some corner houses in cities, or of isolated country residences, the furnace should be so placed as to give short runs of pipe to the rooms on the cold side or sides of the building; in other words, to the *north-west of the centre*, so as to secure short runs of pipe to the rooms on the north and north-west.   Due provision for the north-east rooms must also be made.   In this section of the country the prevailing winds of winter come from the northwest; and the cold, penetrating rain storms from the east and north-east.   These winds tend to force the heated air in the building towards the south-east or south-west rooms, necessitating an ample supply to the rooms from which the warm air is liable thus to be driven.

Two or More Furnaces Often Desirable

In long and narrow buildings, such as the better class of residences in large cities, two furnaces should be used, one to heat the front, and the other the back building. So in general, whenever the use of a single furnace would necessitate a long run of pipe to any part of the building, two or more furnaces are to be advised. A better distribution of heat can always be effected when two furnaces are used, than when only one is employed. An additional advantage in the employment of a second furnace lies in the reserve power thereby afforded in extremely cold weather. A combination hot-air and hot-water apparatus is also a convenient and effective appliance for reaching distant rooms, and for distributing heat evenly throughout buildings in which the use of hot air alone is rendered difficult by peculiarities of construction; as in the case of old houses, in which no provision for hot air flues has been made.

Location of Hot Air Flues and Registers in Dwellings

Hot air flues should never be placed in an outer wall if it is possible to avoid it. Loss of heat and waste of fuel are sure to result. When it is impossible to avoid it, a double tin flue should be used in the wall, with a sufficient air space between the inner and outer flues to economize the heat.

Whenever practicable, the flues leading to upper stories should be entirely independent of the first floor supply. The first floor is the floor that it is difficult to heat properly. Having accomplished that to entire satisfaction, little doubt need be felt as to the successful heating of the upper floors.

In locating the registers on the first floor, it is desirable to place them at the most exposed side of the room to be heated, unless to do so should involve a long run of pipe in the cellar. In that case, better results will be obtained by locating the registers so as to get a short run of pipe with a good elevation. Floor registers are the most effective for use on the first floor, as the hot air rises through them with less interference from wind currents, and a more steady flow is obtained than from wall registers. The objections to floor registers are, the necessity of cutting carpets, and the accumulation of dust, sweepings, etc., which can only be avoided by the exercise of great care.

When wall registers are for these reasons preferred, care should be taken to see that they have register boxes of ample size, and that the flow of hot air to and through the register box is *not checked or impeded* by a narrow inlet. Nothing is more common, in city houses, than to find a large wall register set in hall or parlor, with a register box or casing that has an air supply of not more than 3x8 inches. Such work cannot be satisfactory. If there is a fire-place in the room, or a ventilating register higher than the hot air register, it is well to locate the hot air register on the opposite side, as a better diffusion of heat will thereby be gained before the warm air is withdrawn from the room. Care should be taken, in all cases, not to locate registers where they may interfere with the suitable placing of the furniture of the room.

When these various considerations are comprehended, it is seen how important it is to settle all these matters properly *before the house is built.* It is far more easy and inexpensive to change a *building plan,* than to change a *building.* Architects should make satisfactory heating a primary consideration, and subordinate other details to this.

Hot Air Feed Pipes

These should be of bright charcoal tin, preferably circular in form, either double seamed, or made up with good slip joints lapping not less than 1¼ inches, and well soldered. Sharp turns are to be avoided, and three-piece or four-piece elbows used, where elbows are necessary, in order to diminish friction. Dampers should be placed in each pipe, near the furnace, and marked, by tags or otherwise, to prevent mistakes. For pipes from 10 to 14 inches in diameter, it is desirable to use IX tin. For larger pipes, No. 26 galvanized iron may be used. They should never approach nearer to the joists or ceiling of cellar than 6 inches, and a metal shield should be placed over them when they are nearer than 12 inches.

Vertical Hot Air Pipes

These should be circular in form wherever possible. While flat or oval pipes are commonly used in walls and partitions, such forms increase friction and greatly retard the flow of warm air; and the area of such pipes should therefore be correspondingly increased. Brick flues, unless lined with tin or terra cotta pipe, should not be used for the passage of hot air. The rough interior of a brick flue impedes the movement of the air; and the absorption of heat by the brick walls is very great.

Care should be taken to form the "footing piece" or "starter" of every vertical pipe in such a way as will insure the quick and easy flow of hot air from the feed pipe into the vertical pipe; and also to see that

XII   HINTS ABOUT HEATING

the feed pipe is not pushed into the "footing piece"
so far as to cut off any of the supply. Nothing is
more common, in the "cheap" class of furnace work,
than the blunders just indicated.

In some cities, the law requires that where a hot
air pipe is carried up through the centre of a partition,
the pipe shall be double, with ½ inch or more space
between the two pipes. Where this is not required, it
is possible by exercising care to make quite as safe a
job by using single pipe. Architects and builders
should be careful so to locate partitions and studding,
that the partition pipe can be carried *straight upward*
throughout its entire length. Offsets tend to accumulate heat at the points at which they are used, and
increase risk of fire while impeding the flow of heat.

Partition pipes should be kept 3 inches clear of
studding on each side, and the studding protected by a
tin lining, for which purpose the commonest grade of
tin may be used. Iron laths, or coarse screen wire
should be used across the pipe between the studding,
in place of wooden lath. To sheathe the pipe with
asbestos felt affords additional protection ; and this
should be done whenever the pipe approaches sufficiently near the woodwork of flooring or partition to
occasion the slightest doubt as to perfect safety.

In old houses, which it is for the first time desired
to heat by means of hot air furnaces, and in which the
cutting out of partitions is objected to, hot air pipes are
often carried to upper rooms through closets on the
lower floors. When this is done, the pipes should be
well sheathed with asbestos felt, and all exposed woodwork lined with tin.

Another expedient that is sometimes resorted to
for heating upper rooms for which no encased hot air
flue has been provided, is to carry up a circular pipe *in
a corner* of a lower room. This pipe is then concealed

from view by studding across this corner at an angle of
45 degrees, nailing iron lath or coarse screen wire
across the pipe, between the studding, to receive the
plastering, as in the case of a partition pipe. This
makes a neat finish, and may be used where the cutting
off of the corner is not objected to. The pipe is of
course boxed out in the lower room at the proper
height from the floor, to receive the register; and, in
the upper room, the same finish may be used, or, if
preferred, a floor register may be employed, the latter
method being the least expensive. By placing a partition in the pipe, and boxing out for an additional
register in the adjacent room, it is possible to heat two
rooms on each floor by means of the one pipe. The
pipe should be reduced in size above the register in
lower room, and provided with a hot air damper. Such
a pipe should also be sheathed with asbestos felt.

Whenever a hot air pipe passes through a floor or
a partition, the wood work should be cut away for a
space of at least 3 inches around the pipe, and protected by a double collar of metal for holes for ventilation, or by the use of a soapstone ring, the latter mode
being, in some cities, required by law.

Size of Hot Air Pipes and Registers

It is not practicable, within the compass of this
pamphlet, to lay down rules that shall cover all possible cases. The most elaborate theories often need
modification by practical judgment, based upon experience, before they can be satisfactorily applied.

The requirements of the average dwelling, under
ordinary conditions, are what are herein referred to.

In determining the size of pipes required, the
cubic capacity of the rooms is by no means the only

matter to be considered. The *exposure* is of much
greater importance. Every square foot of glass, every
square foot of exposed wall surface, and every added
possibility of the removal of heat by sharp and penetrating winds, increase the demand for hot air supply;
and this, of course, means that the size of the pipe
used must be proportionately increased. In connection
with our Tables of Furnace Capacities, fuller data will
be given for the determination of these matters.

Generally speaking, the size of pipes used should
be determined with reference to the following considerations :—1. Size of rooms. 2. Exposure. 3. Direction from furnace. 4. Distance from furnace. 5.
Height above furnace ; *i. e.*, whether on first, second
or third floor.

The larger the room, and the greater the exposure,
the larger the pipe required. If the direction of the
room from the furnace is such that the hot air must be
carried to the room against the prevailing winter winds,
the pipe must be *larger* than the pipes used to rooms of
like size on the warm side of the house. So also, a
room that is at a distance from the furnace must have
a larger supply of pipe than a room that is near by, in
order to make up for the diminished elevation of the
pipe.

A room on an upper floor will not require so large
a pipe as one of the same size on the first floor; as the
greater draft of the vertical pipe increases the velocity
and therefore the quantity, of the hot air passing
through it.

As has before been stated, rooms on the first floor
are best heated by independent pipes. Rooms on
second and third floors can usually be heated satisfactorily by single lines of pipe, reduced in size above
second floor register, and furnished with a hot air
damper to regulate the flow to the upper room.

Under ordinary conditions, the sizes of pipes and registers indicated below may be recommended:

FIRST FLOOR

Size of Room in Cubic Feet	Size of Pipe		Size of Register	
	If Round	If Flat	If Round	If Square
Less than 1,500...	7 inches	4 x 9 in.	9 inches	7 x 10 in.
1,500 to 2,000.	8 "	4 x 12 "	10 "	8 x 10 "
2,000 to 3,000......	9 "	4 x 16 "	11 "	8 x 12 "
3,000 to 4,000......	10 "	4 x 18 "	12 "	9 x 14 "
4,000 to 5,500......	11 "	6 x 16 "	14 "	12 x 15 "
5,500 to 7,000......	12 "	6 x 18 "	16 "	14 x 18 "

SECOND AND THIRD FLOORS

USING ONE PIPE, DIMINISHED ABOVE SECOND FLOOR REGISTER

Size of Room in Cubic Feet	Size of Pipe to Second Floor		Size of Diminished Pipe to Third Floor	
	If Round	If Flat	If Round	If Flat
Less than 1,500...	8 inches	4 x 12 in.	6 inches	4 x 9 in.
1,500 to 2,000......	9 "	4 x 16 "	7 "	4 x 9 "
2,000 to 3,000......	10 "	4 x 18 "	8 "	4 x 12 "
3,000 to 4,000......	11 "	6 x 16 "	9 "	4 x 14 "

Size of Room in Cubic Feet	Size of Register—Second Floor	Size of Register—Third Floor
Less than 1,500...	8 x 10 inches	6 x 10 inches
1,500 to 2,000......	8 x 12 "	7 x 10 "
2,000 to 3,000......	9 x 14 "	8 x 10 "
3,000 to 4,000.....	10 x 14 "	9 x 12 "

If the house is but two stories high, use independent pipes to second story rooms, of the sizes indicated in the foregoing tables for diminished pipe to third story rooms, with registers of corresponding size.

In the halls of dwellings, an 8 inch pipe with a
10 inch round or an 8 x 10 inch square register will, in
most cases be found sufficient.

Relative Area of Pipes and Registers

It should always be remembered that the valves
and fret-work of the registers commonly used, reduce
their nominal capacity about one-third.   The following
table of relative areas will be found convenient for
reference:

Hot Air Pipe		Round Registers		Square Registers	
Size	Effective Area	Size	Effective Area	Size	Effective Area
7 in.	38 sq. in.	7 in.	26 sq. in.	6 x 10 in	40 sq. in.
8 "	50 "	8 "	33 "	7 x 10 "	46 "
9 "	63 "	9 "	42 "	8 x 10 "	53 "
10 "	78 "	10 "	52 "	8 x 12 "	64 "
11 "	95 "	—	—	9 x 12 "	72 "
12 "	113 "	12 "	75 "	9 x 14 "	84 "
14 "	153 "	14 "	103 "	10 x 12 "	80 "
16 "	201 "	16 "	134 "	10 x 14 "	93 "
18 "	254 "	18 "	169 "	12 x 15 "	120 "
20 "	314 "	20 "	209 "	14 x 18 "	165 "
22 "	380 "	24 "	301 "	16 x 20 "	213 "
24 "	452 "	30 "	471 "	16 x 24 "	256 "

The tin or galvanized iron register boxes in which
registers are set, should be from one to three inches
deeper, according to size, than the depth of the register
when open.   In setting wall registers in shallow flues,
as in partitions, the register should be set in a stone
border, or else a convex register should be used, so that
the flange and valves of the register may not enter
into and partially shut off the hot air flue.

Churches, Stores and Public Buildings

These structures present somewhat different conditions from those that are encountered in dwelling houses. All that has been said as to the underlying principles of warm air heating of course holds good ; but their application is modified by the circumstances of each case. Systematic and well planned ventilating arrangements are much more frequently found in these buildings than in ordinary dwellings. These serve to facilitate the heating of the building; but they also call for larger heating capacity in the furnaces selected. The mistakes usually made in such cases, are the selection of furnaces that are too small, and the endeavor to make one furnace do the work of two.

In locating registers for church heating, the endeavor should be to distribute the heat *evenly* throughout the building. A register should always be placed *near each entrance*, in order that the effect of the influx of cold air, consequent upon the frequent opening of the doors, may be counteracted. Other registers should be placed wherever they are necessary to carry the heat equally to all parts of the room.

The location of registers having first been determined, the next thing to ascertain is whether these registers can be reached by short runs of pipe, with a good elevation, *from a single furnace*. If not, it may be regarded as settled that more than one furnace will be required.

Try groups of three or four registers, and see whether a point can be found that will give nearly equal, and moderately short, runs of pipe to the registers of each group, and locate furnaces accordingly. Having found the number of furnaces necessary, it will be easy to determine upon their proper size and capacity.

Never locate a register immediately over a furnace.

It is a source of discomfort, to those who sit near it, by reason of the intense heat and strong draft arising from it; while the heat rises rapidly to the ceiling without dispersing its benefits to those who are a little further removed from it. Two or three registers of a smaller size, each located eight or ten feet away from the furnace, will give far more pleasant and satisfactory results.

In arranging the registers for a store, care should be taken to place one *near the entrance.* The location of the others should depend upon the ordinary uses of certain parts of the building. Where sorting, handling and packing of goods is usual, less heat will be needed than in those parts of the building in which persons are engaged in sedentary occupation. In stores in which skylight openings are cut through to the first floor, the first floor registers should be so placed as to prevent the warm air from rising through the opening until after its heat has been well diffused throughout the first floor.

Cold Air Supply

This should never be taken from the cellar if it is possible to avoid doing so; but it should be brought from the outer air, by means of a cold air duct, which may be constructed of brick, galvanized iron or wood, as may be preferred. The sectional area of this duct should be not less than three-fourths of the sectional area of all the hot air pipes leading from the furnace. Thus if four 9-inch pipes are to be supplied with warm air, their total area being 252 square inches, the cold air duct should measure not less than 10 x 19 inches inside, or its equivalent. If one cold air opening in the base of the furnace is inadequate to receive this supply, the duct should be divided into two parts, and one carried to an opening on each side of the furnace base.

Whenever possible, take the cold air from either the north or the west side of the building, as it is from the north-west that the prevailing cold winds of winter come. Put a slide in the cold air duct, arranged so that it can be closed one-half, should an unusual wind-pressure render it necessary, but so that it can never be entirely shut off. The outer opening of the duct should be closed by a wire screen, to prevent the entrance of animals. When a settling chamber and filtration apparatus can be provided, all dust may be removed from the air before its admission to the furnace; but, except in the best jobs of work, the expense of such an appliance occasions objection. Very excellent work can be done if cost is a secondary consideration.

The best method of introducing the cold air to the furnace is *from beneath*. This involves the use of a furnace with closed base and sides, and an open bottom. Place the furnace over a pit, lined with brick laid in cement, first building a central pier up to the ash pit, to support the weight of the furnace. The cold air duct should be so connected with this pit as to secure a perfectly uniform distribution of the cold air around the furnace, in order that the diffusion of heat from the radiating surfaces may be rapid and uniform. When it is not desired to incur this expense, a furnace with closed bottom may be used, and the cold air introduced thereto by means of suitable collars in the sides of the casing.

If it is impracticable to get a direct cold air supply and the air has therefore to be taken from the cellar, the cellar must be kept perfectly clean, and as free from dust as possible; and an inlet for fresh air must be provided by carrying a pipe of the proper size from a window, or an opening in the wall, to a point within twelve or fourteen inches of the cellar floor. The cold air so introduced will flow in a direct line to the furnace, without creating an unpleasant draft in the cellar. Such an expedient, however, should not be resorted to if there is any way of reaching the furnace by a regular cold air duct.

If there are any turns or bends in the cold air duct, care should be taken to avoid any diminution of its area at such points. It must be of *full size throughout.* A furnace cannot supply warm air unless it is first fed with the air that it is expected to heat.

When a public hall or the audience room of a church is to be heated by a hot air furnace, it is sometimes advantageous to make a connection between the cold air duct of the furnace and the room to be heated, arranging it so that this connecting pipe may be entirely closed by a slide. Until the room is occupied by the audience, the cold air may thus be drawn from the room itself and returned to it warmed, the heating process then going on rapidly. As soon as the audience begins to assemble, the connecting pipe from the room should be closed, and the outer cold air supply opened; so that thereafter a supply of pure warm air will be furnished to the room, already comfortably heated.

When it is desired to place a furnace in the basement of a church or other building, and to heat the basement as well as the upper part of the building thereby, the cold air supply should be carried to the furnace beneath the basement floor. To obtain good results, the furnace should be fitted with but a single casing, which should be of Russia iron, in order that the heat may be freely radiated into the basement room. An upper door or doors should be placed in the casing, and a damper in each of the hot air pipes that lead to the room above. The entire heat of the furnace may then, if desired, be retained in the basement by closing the hot air dampers and opening the upper door or doors of the furnace.

Ventilation

In order to remove the carbonic acid gas and organic impurities produced by respiration, and to make good the constant withdrawal of oxygen by the burning of lights at night, some provision for a continual change of the air of inhabited rooms is necessary. This ventilation it is the province of the architect to arrange for; and the furnace setter is rarely consulted. The latter has in most cases to be content with such ventilation as he finds to have been already provided when his own work begins. Yet unless some way is at hand whereby the air that is already in a room may flow out, it is manifest that the hot air which the furnace is ready to supply cannot flow into it. Sometimes the quickest way to heat a room is to lower a window slightly on the side opposite to that from which the wind is blowing, to give the cold air in a room a chance to escape freely, and make room for the admission of the warm air that would otherwise enter but slowly.

To discuss the subject of ventilation at length would require a volume. Only a passing notice, rendered necessary by the intimate connection of ventilation and heating, is possible here.

The use of open fireplaces, as has before been said, so long as fire is kept in them, furnishes to many dwellings a good method of ventilation. When the fire is out, a down draft often occurs in the chimney, which renders it useless as means of removing vitiated air. In large buildings, such as churches and halls, systematic provision is usually made for ventilation; but many dwellings are without suitable arrangements of this sort. The most common method of ventilating dwellings is that of employing outlet flues, which are kept warm either by being built in immediate contact with the smoke flues of the furnace and of the kitchen range, or by having the smoke pipes carried up through the ventilating flues, using for the purpose a pipe made either of cast or wrought iron, or terra cotta. The warmth thus obtained creates an upward current in the ventilating flues, and the vitiated air is drawn out of the rooms and up the flues through registers suitably located and opening into the ventilating flues, either directly or through ventilating pipes.

Some persons argue that ventilating registers should be placed near the floor of the room. They base this opinion upon the fact that carbonic acid gas, *when unmixed*, is heavier than common air at the same temperature; and they therefore contend that when it is produced in a room by respiration, it will fall to the floor, and that it can be removed only by means of outlets at the floor. This notion fails to take into account the law of transfusion of gases, which teaches us that at the moment carbonic acid gas is exhaled from the lungs, it at once intermingles with the air throughout the entire room. It also overlooks the fact that, when it passes out of the lungs, the human breath, loaded with organic impurities as well as with carbonic acid gas, is at the bodily temperature of 98 degrees, while the ordinary temperature of a properly heated room is only about 70 degrees. The heated breath, therefore, rises at once to a level that corresponds with its temperature; so that the foulest air in a room will ordinarily be found at a higher level than the heads of its occupants. If any one doubts this, let him simply stand upon a table in a heated room of ordinary height, and find whether the air that he will then inhale is purer and sweeter than the air he was breathing when he stood upon the floor.

Observation leads to the belief that in ordinary dwellings the most satisfactory results are attained when the ventilating registers are placed near the ceiling. This plan, of course, continually withdraws heat from the room, and demands an ample supply of hot air, larger furnaces and more fuel. Like almost every other good thing, good ventilation costs money. When economy of fuel is an object, place the ventilating registers near the floor. Architects often provide outlet registers near the floor and near the ceiling also, leaving the occupants of the house free to open either at their pleasure.

Ample Furnace Capacity Essential

We repeat that wherever means of artificial ventilation have been provided, the furnace should be of ample capacity, otherwise the rooms may be cold when the ventilating registers are open; and if they are not to be opened, they might as well not exist at all.

Fortunately in ordinary dwellings, tenanted, as most of our American homes are, by but a few persons, natural ventilation furnishes all the change of air that is indispensable to health, if the rooms are heated by a good hot air furnace, well supplied by cold air from without. The pure warm air that enters the room from the furnace is *constantly displacing an equal amount* of the air that was previously in the room. If this were not so, the warm air could not enter the room at all.

This displacement is made possible by the outlet that is afforded by crevices in floors and around window frames, and by loosely fitted doors and window sashes, and lastly, though not least, by diffusion through the walls themselves. This has been shown, by Pettenkofer's experiments, to be not less than seven cubic feet of air per hour for each square yard of wall surface (brick wall, plastered, but not papered), when the difference between the temperature within and without is 40 degrees. In a room 12 x 15 x 10 feet, this diffusion would amount to 2,800 cubic feet per hour.

As has before been said, ordinary dwellings are large in proportion to the number of persons who live in them; and natural ventilation is often adequate to effect the necessary change of air. In the light of what has been said, however, the great importance of a plentiful supply of pure air to the furnace must clearly appear. In some States the laws require 1,800 cubic feet of air to be supplied in school rooms per hour for each scholar. This is a very fair and healthful standard where rooms are occupied for two or more hours at a time. In ordinary hospital service 3,600 cubic feet per patient per hour, and for infectious diseases, double that amount should be provided. It is not that such an amount of air is required for the simple act of breathing, but that the emanations thrown off through the pores vitiate the air.

These considerations also show that it is short-sighted economy to stint the size of the furnace used. No matter whether reliance is placed upon natural or artificial ventilation, *ample furnace power* must be provided if a steady and adequate change of air in the rooms is to be secured.

Supply of Moisture

This is a matter of some importance. As air is heated, its capacity for absorbing moisture proportionately increases. If there be no arrangement made for supplying this moisture directly to the air as it is heated, it will be drawn from the wood work and furniture in the house, causing annoying and damaging cracks and shrinkage. The health of the occupants of the room also will suffer, as the needed moisture will be taken up by the heated air from the bodily surfaces and the mucous membranes, thereby rendering the persons susceptible to cold, and occasioning many catarrhal troubles. In all our furnaces provision is made for a water supply; and the pans provided for that purpose should always be kept filled with water.

Hot Water and Hot Air Heating Combined

Difficulties often arise in properly distributing furn·ace heat in buildings in which a small number of rooms are too far distant from the furnace to be properly heated, or in which the carrying of horizontal hot air pipes through finished rooms is objectionable. To overcome such difficulties, we manufacture a furnace with a water-heating attachment by which these rooms can be heated without increasing the number of fires, or using an additional furnace, or disfiguring finished rooms with large hot air pipes. As the problems involved are practically those involved in heating by hot water, it becomes necessary to present a few explanations and rules which are intended to be plain, simple and easily understood, without going too much into detail. In accomplishing this, several considerations require attention, namely:

FIRST.—(*a*) The use of the rooms, and the number of people who are to occupy them.

(*b*) The cubic contents, the exposure and the glass surface.

SECOND.—(*a*) The kind of radiation, whether direct or indirect.

(*b*) The amount of radiation required.

(*c*) If indirect, the proper size of air ducts.

(*d*) The location and arrangement of radiators

THIRD.—(*a*) The arrangement of flow, return and draw-off pipes.

(*b*) The proper size of same.

(*c*) The location of expansion tank with overflow and supply pipes for same.

FOURTH.—The proportion of water-heating surface to the amount of radiating surface, and the size of hot air furnace required in connection with the hot water heating.

Taking these up in order, we consider:

FIRST.—(*a*) The use of the rooms and number of occupants.

In an ordinary dwelling house, there are two rooms that should be especially well provided with heat—the dining-room and the bath-room. The dining-room, in which persons first assemble in the morning, after the fire has run low during the night, should be so arranged as to be heated quickly and well, as breakfast in a cold, cheerless room is not conducive to a pleasant disposition during the remainder of the day. The bath room requires much more heat in proportion to its size than other rooms, as it is generally a small room, and the smaller the room the larger must be the proportion of heating surface. Direct radiators are more comfortable in rooms of this kind; for a current of air from a register, even though quite warm, seems much cooler than it really is, especially if one is unclothed and wet. In ordinary living rooms, it is sufficient both for proper heating and ventilation to change the air in the room twice per hour. In rooms in which large numbers of people congregate, a proportionately larger supply of air is needed, but the volume of air delivered should be of a much lower temperature, as the heat radiated from each person, as well as that thrown off from the lungs, not only tends to vitiate the air, but so raises the temperature that, in ordinary winter weather, the bodily heat of the occupants will maintain the heat of the room, and the air supply should then only be heated sufficiently to take the chill off.

(*b*) Cubic contents, exposure and glass surface. These are the main factors to be considered when the rooms are used as living rooms, offices, etc., to be occupied by but few persons at the same time. Specially exposed rooms are those situated on the north or west side; as in the winter season they get little or no sun rays, and are exposed to the colder winds that force the heat from those rooms to the opposite side of the building; also corner rooms or others that have two or three outside exposures. Rooms with but one outside exposure are classed as ordinary rooms.

SECOND.—(*a*) THE KIND OF RADIATION. "Direct" and "Indirect," are terms used to designate the location of radiating surfaces and the manner of supplying or conducting heat to the room. A "direct" radiator is one that is located within the room or space to be heated, communicating its heat directly to the air that is contained in the room.

An "indirect" radiator is one that is located at some point beneath the room to be heated, being encased in a galvanized iron casing or box to which

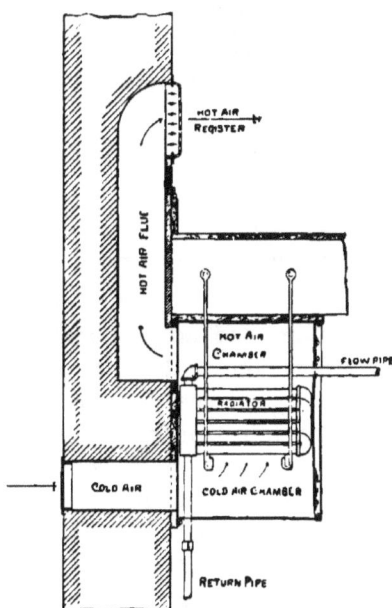

CUT OF INDIRECT RADIATOR.

fresh air is supplied, and from which the heated air thus supplied rises through suitably arranged ducts to the room that is to be warmed.

A radiator located within the room to be heated, that is supplied with external air in such a way as to heat it before it passes into the room is called a "semi-direct," or "direct-indirect" radiator. This style of radiator is growing in favor, as it combines the advantages both of direct and of indirect radiation.

(*b*) AMOUNT OF RADIATION REQUIRED. Ordinarily, on account of the constant change of air, from 45 to 50 per cent. more of indirect radiating surface and about 30 per cent. more of "semi-direct" radiating surface is required than of direct radiating surface, to do the same amount of work.

SEMI-DIRECT RADIATOR

COLD AIR

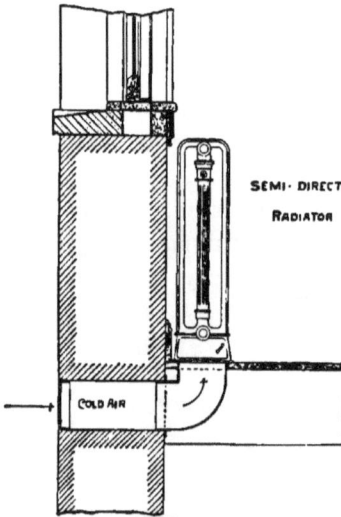

CUT OF SEMI-DIRECT RADIATOR.

For rooms of ordinary exposure, *i. e.*, with but one side exposed to the outer air, and an ordinary amount of glass surface, a proportion of three square feet of direct hot water radiating surface per hundred cubic feet of space, is a very fair standard; for exposed rooms, from four to four and a half square feet, according to the degree of exposure; in determining which, good judgment must be used, as no "hard and fast rule" will strictly apply to all cases. Upon this basis, we may compute as follows:

CONTENTS OF ROOM	DIRECT RADIATION FOR ORDINARY EXPOSURE	DIRECT RADIATION FOR SPECIAL EXPOSURE
1,000 cubic feet.	30 square feet.	40 to 45 square feet.
1,500 " "	45 " "	60 " 67 " "
2,000 " "	60 " "	80 " 90 " "
2,500 " "	75 " "	100 " 112 " "
5,000 " "	150 " "	200 " 225 " "
10,000 " "	300 " "	400 " 450 " "

Having proportioned the amount of direct radiation required, add thereto 30 per cent. of radiation, if semidirect or "direct-indirect" radiation is to be employed. If the full indirect system is to be employed, add 50 per cent.

To illustrate: a room 12 feet wide, 15 feet long and 10 feet high, containing 1,800 feet, if of ordinary exposure, will require 54 feet of direct radiating surface; of semi-direct radiation, 30 per cent. more, or $70\frac{2}{10}$ square feet; or of indirect radiation, 50 per cent. more, or 81 square feet. A room of the same size, specially exposed, will require from 72 to 81 square feet of direct radiating surface; from $93\frac{6}{10}$ to $105\frac{3}{10}$ square feet of semi-direct radiation; or from 108 to $121\frac{1}{2}$ square feet of indirect radiation.

The following is also a convenient working formula for computing Hot Water Radiation :

1. Divide cubic feet of air in room by 75.
2. Add to quotient the actual square feet of glass in room, measuring between casings.
3. Divide square feet of exposed wall by 10, if wall is from 8 inches to 13 inches thick or by 15 if wall is more than 13 inches in thickness, and add the quotient to the above sum.
4. Multiply the sum total by .70 if for direct radiation, by .95 for semi-direct radiation ; or by 1.05 for indirect radiation.

For example : A room 10x2x100 feet is exposed on two sides, has three windows each 3x5 feet, and wall 8 inches thick. To find radiation required :

$$10 \times 20 \times 10 = 2{,}000 \text{ cub. ft.} \div 75 = 26.6$$

3 windows each 3x5 = 45 sq. ft. glass 45.
30 lineal feet wall x10 ft. in height=300 sq. ft.

Less glass surface 45.

$$255 \div 10 = 25.5$$

Total 97.1

$$97.1 \times .70 = 67.9 \text{ sq. feet direct radiation}$$
$$97.1 \times .95 = 92.2 \text{ " " semi-direct "}$$
$$97.1 \times 1.03 = 101.9 \text{ " " indirect "}$$

(*c*) THE PROPER SIZE OF AIR-DUCTS, when indirect radiation is used. These should be proportioned in accordance with the purposes for which the several rooms are intended to be used. Rooms on the first floor require larger ducts than those located on upper floors, as the greater the vertical height of the air-duct the greater will be the velocity of the flow of air through the duct. For an ordinary living room, as has been said before, it is sufficient for both heat and ventilation if the warm-air supply is large enough in volume to change completely the air in the room every thirty minutes. The air in a crowded room, however, should be changed every fifteen minutes ; the volume of air thus supplied being at a lower temperature, as previously stated. Elaborate calculations as to the volume of fresh air required per occupant, in crowded rooms, are unnecessary here ; inasmuch as there is great difference of opinion among competent authorities upon this point, and as it is practically impossible to change the air completely, in any large room, oftener than once every fifteen minutes without forced ventilation.

The velocity of warm air in a vertical duct varies with the height of the duct, and the difference between the external temperature and that of the air in the flue. For practical purposes, under average conditions, a duct 144 square inches in sectional area will deliver 10,000 cubic feet per hour to a room on the first floor; while one of 120 square inches sectional area will readily deliver the same quantity per hour to the second floor, and one of 96 square inches sectional area to the third floor, of a building of ordinary height. Thus, if a crowded room is to be heated, having a capacity of 10,000 cubic feet, the air in the room should be changed 4 times per hour, and the combined sectional area of all hot-air ducts leading to it from indirect radiators should be 144x4=576 square inches, if the room is on the first floor ; 120x4=480 square inches, if on the second floor ; or 96x4=384 square inches if on the third floor. If the room is occupied simply for ordinary living purposes, one-half this flue area will suffice. This simple formula will furnish a ready means of calculating the sectional area of any warm air ducts, under average conditions.

The cold air supply ducts should be of not less than three-fourths the area of the exit or vent ducts, for the reason that when the full volume of air is admitted, it is but slightly heated, or, as heretofore expressed, "the chill taken off." Mistakes are often made in not making the fresh air supply to indirect radiators large enough. The supply ducts are calculated upon the basis of a quiet or still air, the movement of the air being caused by the heat of the furnace. The air can be called quiet when moving not over one mile per hour, or about 1½ feet per second, which is called an "imperceptible breeze"; and this condition often occurs in clear, cold weather. When the air is still, all cold air ducts should be fully opened. When the wind is blowing at the rate of six miles or upward per hour, directly into the cold air duct, the supply of air to the radiators should be regulated by means of a properly fitted damper, which should always be so placed in every cold air supply duct as to be conveniently reached.

(*d*) LOCATION AND ARRANGEMENT OF RADIATORS. Direct radiators should be placed in, or as near to, the colder parts of the room as possible. Semi-direct radiators are preferably located next the outer wall, under the windows, so that fresh air can easily be conveyed to them. Indirect radiators should be placed as near the uptake, or vertical flue, as possible. If more than one uptake is arranged from a stack of indirect radiators, the stack should occupy a position as nearly as possible central between the flues, giving preference, however to the flues that are nearer the prevailing cold winds; it being remembered that the tendency of air in the rooms is in the same direction as on the outside, and that when a strong wind prevails, it is difficult to carry the hot air against the wind much more than twelve feet horizontally.

THIRD. (*a*) THE ARRANGEMENT OF FLOW PIPES, &c. In heating with hot water, it should be carefully noted, that in filling the apparatus, from the bottom upward, all the air will pass out at the highest points. There should be no air pockets whatever in the system. The flow pipes should incline upwards from the water heater to the riser pipes or radiators on or above the first floor; and the return pipe should incline downwards, and may be carried underneath the floor level on which the furnace is placed rising at the furnace to the water heater. For indirect or direct radiators that are located on the same floor as the furnace, the flow connection pipes should pitch downward toward them, so that the air when they are filled will pass upward toward the higher points to be let off at some higher radiator or riser. All direct radiators above the furnace level should have air cocks at the highest point of radiator, to relieve them of air that may accumulate in filling the apparatus, or that may be freed from the water afterward. At points at which a reduction of all the sizes of pipes occurs, or where a larger pipe is reduced to a smaller, or where branches are taken off for radiators or risers above, eccentric tees should in all cases be used to leave the pipes fair with each other on the top, so that no air can accumulate at these points.

(b) SIZES OF PIPES. The flow and return pipes should be of the same size. The area of larger pipes that supply smaller ones should equal the combined area of the smaller ones. Connecting pipes to radiators should be as follows:

```
For 50 sq. ft. or less of radiating surface,  ¾ in.
 "  50  "   "  to 80 square feet.......... 1    "
 "  80  "   "  to 150  "       "   ......... 1¼ "
From 150 "  "  to 200  "       "   ......... 1½ "
 "  200  "   "  to 350  "       "   ......... 2    "
```

Smaller than ¾ inch pipes are not advisable, on account of their liability to get clogged by sediment. In long runs to radiators, the pipes may have to be increased in size on account of the loss of speed to the current by the friction of the water in the pipes.

(c) The expansion tank should be located not less than four feet above the top of the highest radiator, and the supply pipe from the tank to the heating apparatus should be preferably connected directly with the water heater. It may, however, be connected with a return riser, upon which there are no valves between the tank and the water heater. It is well where a return riser is used to connect the corresponding flow pipe riser to the return riser. If a separate supply from the tank to the apparatus is used, the supply should be connected to the return pipe near the water heater, and also connected to the top of the flow main at the ceiling directly over the furnace. The water supply to the tank should have a ball cock to keep the expansion tank constantly filled. An overflow pipe of at least four times the area of the tank supply should be carried into a water closet cistern, or to some other suitable place to discharge the surplus of water as it is expanded by the heat. It should never be directly connected with the soil pipes, as there is no expansion and contraction of the water when the apparatus is not in use, and the water in the trap would evaporate and admit the sewer gas through the overflow pipes. In situations where there is no head of water to reach the expansion tank, and the apparatus has to be filled by pump or otherwise, an expansion tank of sufficient size to receive the whole expansion of the water can be used without wasting any water. In such a case the tank should be a little larger than one twenty-fifth of the contents of the entire apparatus, as the water from ordinary temperatures heated to the boiling point expands about one gallon in twenty-five.

FOURTH.—The proper size of the water heater depends very much on two factors, viz: The amount of hot water radiating surface, and the amount of heat required from the hot air part of the apparatus. If the hot air furnace is small, a much "hotter fire" has to be carried to heat the rooms, and a smaller water heater is required than where the furnace is large and amply sufficient to heat the rooms with a slow and vastly more economical fire. It is well to remark here, that the main point to be considered in installing a heating apparatus is not what it costs to put in the apparatus, but what it costs to run it.

A large furnace that will not ordinarily require feeding more than twice during the 24 hours is much more economical in fuel, will last longer, require less labor, and be found altogether more healthful and satisfactory than a small one which requires constant attention.

To attain satisfactory results from a combination heating apparatus, the balance between the amount of space to be heated by hot water and that to be heated by hot air must be carefully observed when installing the apparatus. When proper proportions are employed, the parts of the building heated by hot water will be equally heated with those parts that are heated by warm air, and there will be no generation of steam nor boiling of water in the expansion tank. The following table gives the proper relative heating capacities of the Paragon Combination Furnaces, determined by actual experience. It is based upon direct radiation, mains being uncovered. If mains are covered, from 15 to 20 per cent. additional capacity in radiation will be gained. If the radiation required in the rooms to be heated by hot water is less than the total amount given in the table, the proper balance should be secured by placing a radiator in the hall of the building, sufficiently large to make up the full heating capacity; or by putting a radiator or two in rooms that are also warmed by hot air, and by placing a large hot air register in the hall, which can be opened should the furnace at any time appear to be over-heated. When good judgment is exercised in these respects, a perfect balance is maintained and a thoroughly equal temperature secured in in all parts of the building.

PARAGON COMBINATION FURNACE

Size	Diameter of Fire Pot	Heating Capacity in Square Feet of Direct Radiation	Heating Capacity in Cubic Feet by Hot Water	Heating Capacity in Cubic Feet by Hot Air when Divided into Rooms as in Residences
40 in. casing	23 inches	400	12,000 to 14,000	10,000 to 12,000
44 " "	25 "	450	14,000 to 16,000	12,000 to 15,000
48 " "	28 "	525	16,000 to 22,000	15,000 to 18,000

Diameter of Flow Pipe on 40 and 44 inch sizes, 2½ inches.
" " " " " 48 inch size, 3 inches.
The return pipe or pipes should equal in diameter the main flow pipes.

There should be placed on the flow pipe near the furnace, a thermometer to indicate the temperature of the water, which should never be above 210° Fahrenheit. At the lowest point the return pipe should have a valve to draw the water off the entire apparatus to prevent freezing should the house be vacant in cold weather.

In the latter part of Part II. of this book, will be found cuts representing floor plans of a long and narrow city house heated by a Paragon Combination Furnace, which fully illustrate the application of the principles that govern successful work of this character.

PART TWO

HINTS ⚜ ABOUT HEATING

Containing valuable suggestions respecting hot-air furnace work, together
with tables of dimensions, capacities,
etc., prepared with especial reference
to the Paragon Hot Air Furnace

Part Two ❧ Third Edition ❧ Revised and Enlarged

Philadelphia and Baltimore

ISAAC A. SHEPPARD & CO

Mdcccxcviii

Press of
Charles Austin Bates
New York

HINTS ABOUT HEATING

PART TWO

T HE former part of this treatise deals with technical matters referring to the problems that are involved in the proper distribution of heated air, and that are of interest chiefly to architects and builders. Any one specially desirous of obtaining it will receive a copy on application. The following pages, however, contain all that the purchaser of a furnace specially needs to know.

Introductory Remarks

Without touching upon technicalities, we venture to condense within a brief space a few general statements:

Proper furnace heating is perfectly healthful, safe and economical.

Its first cost is less, its management is more easy, and its repair cost is less than that of heating by steam or hot water.

A good furnace, properly installed, does not throw out either gas or dust.

There is no difficulty in evenly heating a building by means of hot air furnaces, if they are put in by one who understands the principles of heating by hot air, and who is adequately paid. Good work cannot be done for a merely nominal sum.

One furnace should not be expected to do the work of two; nor should a small furnace be set to do the work of a large one.

Hot air pipes and registers must be of sufficient size and judiciously located.

A furnace cannot work without a supply of pure air at the bottom, any more than a pump can bring water up from a dry well. This air supply should come from outside the building.

Ample supply of warm air means good ventilation. Every cubic foot of pure fresh air that enters a room pushes a cubic foot of vitiated air out of it. For this reason avoid small furnaces. Be sure that your furnace is big enough.

Keep the water-pan of your furnace full, and you will never suffer from dry, parched air. Neither will your furniture shrink and crack because of a deficiency of moisture in the air.

Use good coal, of a suitable size, if you want a good fire, and plenty of warmth. The heat is not generated *by the furnace.* No furnace can give out more than goes into it. Poor coal—little heat.

A good furnace may be *badly managed.* It has been said that the 'Lord sends food, but that cooks are often sent by'——well, by a very different personage. A stupid servant may ruin a good furnace quite as easily as a good dinner. It pays to instruct servants in furnace management.

The Production of Heat

The problems that are involved in the generation of heat consist of matters that relate to furnace *connections*, furnace *construction*, and furnace *management*. Of these matters, every householder should possess some information. Under the head of furnace connections the smoke pipe and chimney call for consideration.

The Smoke Pipe

This is best made of heavy galvanized iron, well riveted, each section entering the next by a lap of not less than one and one-half inches. The size should be the same throughout as that of the pipe collar of the furnace, and it should run as directly as possible from furnace to chimney, with a steady ascent all the way. Where turns in the pipe are unavoidable, three-piece or four-piece elbows should be used. If the pipe is long and the cellar cold, it will be well to wrap the pipe with asbestos sheathing to prevent loss of heat, which results in impaired draft.

It is well to rivet a flange or collar to the pipe some five inches from the end that enters the chimney. This will prevent the pipe from being pushed at any time too far into the chimney, and will also serve to prevent the leakage of air into the chimney around the pipe. The pipe hole in chimney should be made to fit the pipe neatly. The pipe should be securely wired to the chimney to prevent displacement, and supported throughout its entire length by strong wiring to joists at proper intervals. Screw-hooks are better to wire to than nails, and make a neater finish.

If the smoke pipe has to pass through any partitions, double collars of metal should be used around it with a space of three or four inches between them, this space being ventilated by ample perforations. The pipe should be kept as far as possible from any exposed wood work, and the wood work protected by asbestos sheathing or bright tin, or both, according to the relative position and nearness of the pipe.

The Chimney

The chimney with which a furnace is connected is
a matter of great importance. "Draft," as it is called,
is a function of the *chimney*, not of the furnace. The
upward movement of air in the chimney is due to the
difference in weight between the warm air in the chimney and the cold air outside. The more nearly equal
the temperature within and without the chimney-shaft
the weaker the "draft;" and, *vice versa*, the greater the
difference of temperature the stronger the draft. The
longer the column of air in the chimney, the stronger
will be the draft; so that, other things being equal, the
taller the chimney, the more powerful will be the movement of heated air within it. Leakage of air into the
shaft at any point diminishes the upward pressure; and
if the inside is rough, the draft will be impeded by the
friction of the chimney walls.

Poor chimneys occasion much trouble; and the
difficulties that are due to their imperfect construction
are often the source of complaints respecting the opera-
tion of furnaces. Chimneys should always be built in
the *inner walls* of houses, where possible. If they must
be built in exposed outer walls, let the wall selected be
a south or east wall, and not one on the north or west
side. The chimney should be of adequate size for the
work required of it, but *not too large*. For ordinary purposes, a round flue of smooth terra cotta or tile, of 8
inches inside diameter, is the best. A flue 8 x 8 or 8 x
12 inches in the clear, smoothly pargetted with good
mortar, however, will be found to give good results, if
of proper height.

Chimneys should, if possible, be topped out *above
the highest point* of the roof of the building, in order that
the wind, in passing over the roof, may not occasion
downward currents in the flues and impair or destroy

the draft. A clean-out door should always be located at the base of the chimney; and the bricklayer should always leave the chimney clear of any mortar or other debris.

If hot air flues are built in chimney adjoining the smoke flue, they should be well lined with tin, and the intervening wall well built and carefully pargetted to prevent leakage of gas into the hot air flues.

Before connecting a furnace with the chimney, the chimney should be carefully examined, and cleared of all accumulation of soot or other obstructions, any cracks in chimney stopped, and all unused pipe holes tightly closed.

A Few Words About Gas

It should always be remembered that it is upon *the conditions of smoke pipe and chimney* that freedom from gas depends. Combustion generates gases that will find their way out from the furnace by the channel that offers the *least resistance*. If the draft of the chimney is good and the smoke pipe unobstructed, they will readily pass out into the chimney. Under such conditions, the air pressure upon the furnace is *from without, inwards;* and even if there should be any defects in the joints of the furnace (a thing which after long use may possibly occur), air will be carried through such a defective joint, *into* the furnace, instead of gas passing out through it.

If, however, the outlet into the chimney be so impeded, or the draft of the chimney so defective, that the gas finds *less resistance* in passing out through the joints of the furnace than through the smoke pipe and chimney, it will seek an outlet through the joints into the air chamber; or, if the furnace joints are absolutely gas tight, it will pass into the cellar through the doors of the furnace. A good flue, ample connections and a steady fire, afford the surest guarantee of freedom from gaseous products.

Furnace Construction

A well constructed furnace is one that combines simplicity and case of management with durability, freedom from gas and dust, and large radiating surface in proportion to the area of the surface of the grate. These essentials having been first secured, compactness of form and economy of first cost are to be sought for. We know of no other hot air furnace that so fully meets all these requirements as the PARAGON FURNACE; although we also make other excellent goods of this sort, which maintain a deserved popularity, and which are fully equal in efficiency to many furnaces which are erroneously claimed by their manufacturers to be "as good as the PARAGON."

Furnace Management

Specific directions for the successful use of particular furnaces will be given in their proper place. Some general instructions, applicable to all alike, may be given in a few words.

The coal used should be of good quality, and *not too large*. The proper sizes of anthracite coal are a *medium stove size*, for furnaces of moderate capacity, and *large stove size*, or "*egg*" coal, for furnaces with 40 inch casing and upwards. The so-called "white ash" coals give more heat than the "red ash," but require a stronger draught for complete combustion. Where the draft is good, the "white ash" coal is to be preferred.

Pea coal is a size that is sometimes used. It is not what is called a "prepared coal," but it is the largest size sifted out of refuse coal, after the "prepared" sizes, the smallest of which is "chestnut," have been picked

HINTS ABOUT HEATING 9

out. It contains much small slate and dirt, and, unless burned in a shallow fire, cakes and clinkers badly, requiring frequent attention, and diminishing the effectiveness of the furnace.

The fire chamber *must be kept clear*, any accumulations of ashes or clinker being removed as fast as they form. Ashes and clinker have *no heating power*.

"No heat without fuel." The fire chamber *must be kept full* if the house is to be kept warm. A few inches depth of coal upon the grate is insufficient.

A *moderate but steady fire* should be kept. Less clinker will be produced, less wear upon the furnace occasioned, and less coal consumed than by alternately letting the fire burn violently, and then suddenly checking it. Irregular firing burns out furnaces and wastes fuel.

After fresh coal is put on the fire, it should always *be allowed to burn up a little* until the fresh coal is heated through. This prevents the chilling of the fire and causes the gas that arises from the fresh fuel to pass freely into the chimney.

Ashes should be entirely removed from the ash pit at least *once in every twenty-four hours*. Ashes left under the grate impair the draft of the furnace, and *cause the grate to burn out*. It is cheaper to attend to this *than to buy new grates*.

Carefully study the varying draft of the flue with which the furnace is connected, and regulate the furnace in the way which experience demonstrates is best suited to the conditions under which it operates. These differ in almost every case, and can be determined only by close observation.

If the house is to be comfortable in the morning, the furnace must be so regulated in the evening as to keep the temperature of the first floor rooms from falling too low during the night.

Intelligence, observation and patience are necessary
to manage properly any form of heating apparatus. The
exercise of these qualities sometimes fails in the case
of servants, to whom the management of furnaces is
ordinarily entrusted. In such instances some member
of the family should supplement the deficiencies of the
person who has the care of the furnace.

Importance of Proper Plans

What has been said respecting the location of hot
air flues and registers, emphasizes the importance of
planning suitable arrangements for the house-heating
before the house is erected. It is not only unfair to the
furnace setter, but a detriment for all time to the occupants of the house, to build it without carefully arranged
and suitable provision for a proper distribution of hot
air throughout the dwelling. Many an architect, sound
as his professional judgment may be regarding most
matters, would find it of great advantage to submit his
plans for hot air work to a skilled and intelligent fur-
naceman before their completion. He will then find
that when the time comes to set the furnace in place, it
will not be necessary either to put up with unsatisfactory
results or to make expensive and annoying alterations
in the building.

In cases in which it is the intention to specify furnaces of our manufacture in new buildings, we shall be
glad, so far as our engagements will permit, to confer
with architects or builders respecting these matters,
and to make any suggestion that may aid them in
obtaining the best results.

False Economy

A word of caution here to owners of property:

There are all sorts and sizes of furnaces, and all
sorts and kinds of furnace work. The poorest is cheap

enough. That which is really good cannot be had without paying for it what it is worth. The man who flatters himself that he is getting *more than what he pays for* is grievously mistaken. If the work of setting a furnace is slighted, the furnace will be overtaxed, and it will soon burn out. So also, if the furnace be of poor quality, or if it be too small to do the work required of it, it will not last long, and a new one will soon have to be purchased. Meanwhile, the occupants of the building will be more or less inconvenienced, and perhaps injured in health. Such attempted saving is *false economy*. To buy a furnace of good quality and of ample capacity, and to have all the work connected with it properly arranged and put up of good material, in a thoroughly workmanlike manner, will be found the most satisfactory, as well as the cheapest plan, in the end.

What We Need to Know

When we are asked for information or advice respecting furnace work, or the selection of a furnace, the following information should be given us:

I. Is the building constructed of brick, stone or wood?

II. Is it one of a block, or does it stand alone?

III. If alone, is it much exposed? Give particulars.

IV. Draw a plan, no matter how rough, of the cellar and of each floor above. Mark dimensions of each room. State height of ceilings of each story; and the height of cellar clear of joists. Mark location and size of smoke flues and of hot air flues; also any open fire-places, and any closets or recesses through which hot air pipes may be run if necessary. Mark also the location preferred for each register.

V. Mark the points of the compass on the plan.

VI. If any girders in cellar, mark them on plan and state their clear height above cellar floor. Mark also any piers or other obstructions to the run of hot air pipes in cellar.

VII. State whether there are any objections or difficulties to interfere with digging a pit in cellar to lower the furnace, if necessary to do so, in order to give a better elevation to the hot air pipes.

VIII. Mark on the plan the cellar window or other opening through which the cold air supply is to be taken. Remember that this should be on north or west side of building.

IX. If a church or other public building, mark on plan the location of doors, windows, vestibules, aisles, pulpit, etc., also ventilating flues, if any. Also state whether there are any open spaces under pews for circulation of air.

X. State whether the building is still to be constructed, or whether it has already been completed.

NOTE.—If the building has not yet been erected, the details of heating plan should be settled without delay, in order that suitable provision for a good job of work may be made as the structure is built. Proper plans insure the comfort of the occupants, as well as economy of fuel and durability of heating apparatus.

We Furnish Blanks

We gladly furnish blanks on application, upon which plans may be sketched and necessary questions answered. When such a blank is filled up and mailed to us, the sender may rely upon receiving prompt and accurate information. We make no charge for such service.

PARAGON HOT AIR FURNACES

WITH EQUALIZED DRAFT

Patented August 5, 1890 ❧ Improved 1897

Cut shows Finished Furnace with Draw Centre Grate and Water
Pan at Side ❧ Made also with Water Pan in
front, as shown on following pages

ISAAC A. SHEPPARD & CO.,

Patentees and Manufacturers

Fourth St. and Montgomery Ave.   Eastern Ave. and Chester St.
Philadelphia, Pa.   Baltimore, Md.

Paragon Hot Air Furnaces
With Equalized Draft
Patented August 5, 1890.   Improved 1897.

These furnaces embody the latest and most desirable improvements that modern ingenuity has suggested. Since the patent upon this construction, embracing five distinct specifications, was granted, the PARAGON FURNACE has attracted an unusual amount of attention and has attained conspicuous success. The characteristic features of the PARAGON FURNACE will be seen upon examining the engravings upon the pages that follow. Much has been done during the seven years in which the PARAGON has won its way to the very front rank, to perfect details of construction and to adapt it to every possible use to which a hot air furnace may properly be put, as well as to add beauty of form and design to the structure; but the great principle of the "EQUALIZED DRAFT" has not been improved upon, nor is it easy to see how it can be.

What is Meant by "Equalized Draft"

When we say that PARAGON FURNACES have a "perfectly equalized draft," we mean that in this furnace is embodied, for the first time, the principle of compelling the draft to pass through the whole mass of fuel with *equal force at all points*, the combustion being absolutely even at all parts of the fire pot, from centre to circumference. The products of combustion divide into two currents, one passing outward into the outer drum, and the other upward through the inner drum into the upper radiator, and thence through the connecting arms into the outer drum, where both currents unite, and pass together through the smoke outlet, where the check draft regulates perfectly the draft of the chimney. The equalized draft is the cause of the large saving of fuel effected by the PARAGON, as compared with all other furnaces. It also results in leaving but a small residue of ashes, the fuel value of the combustible being perfectly utilized. The inner drum has a constant current of heat through it, which no other three-drum furnace admits of, the live heat in this drum being especially effective; and the wear upon all the drum casings is equally distributed, increasing their durability.

Beauty of Design

After the first cost of the pattern has been borne, it is just as easy to make an artistic casting as to make an ugly one. The design of the PARAGON FURNACE front is the work of an accomplished sculptor. The accompanying illustration of the hand-door in the upper front gives a taste of its quality. The manufacturers have spent money to beautify the exterior of the PARAGON, because they believe the excellence of its interior construction deserves a handsome dress. They thus set up a high standard of finish before the artisans in their employ; and the result is that the workmen take more pride in their work, and try to bring all parts of the furnace to a more careful fit and finish. The PARAGON FURNACE is to-day the most perfect example of the furnace-moulder's skill that can be found anywhere.

The PARAGON FURNACES are made in a variety of

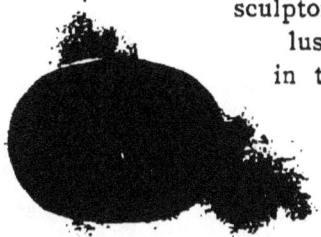

forms, and in several sizes, adapting them to the requirements of different localities, and the taste and needs of different purchasers, namely:

With Sheet Steel Radiator

Six sizes portable form; Three sizes brick set form.

With Cast Iron Radiator

Six sizes portable form; Three sizes brick set form.

Combination (Hot Water and Hot Air)

Three sizes portable form; Three sizes brick set form.

With Sheet Steel Radiator ·

SIX SIZES:

No. 328–A. With High Front; Diameter of Outer Casing, 28 inches; Diameter of Fire Pot, 16 inches.

No. 333–A. With High Front; Diameter of Outer Casing, 33 inches; Diameter of Fire Pot, 19 inches.

No. 336–A. With High Front; Diameter of Outer Casing, 36 inches; Diameter of Fire Pot, 21 inches.

No. 340–A. With High Front; Diameter of Outer Casing, 40 inches; Diameter of Fire Pot, 23 inches.

No. 344–A. With High Front; Diameter of Outer Casing, 44 inches; Diameter of Fire Pot, 25 inches.

No. 348–A. With High Front; Diameter of Outer Casing, 48 inches; Diameter of Fire Pot, 28 inches.

Constructive Features of Steel Plate Furnaces

Upon a strong and roomy ash pit is placed a heavy, corrugated fire pot, the joint being arranged to pack with sand or cement, to make it perfectly gas tight. The fire pot is surmounted by a heavy casting known as the lower radiator, which is cast in one piece. This is carefully proportioned in thickness, and strengthened throughout by corrugation, to prevent cracking by fire. The joint between the lower radiator and the fire pot is also a sand joint.

Three heavy steel plate drum-casings are accurately fitted to flanges cast upon the upper surface of the lower radiator.

PARAGON HOT AIR FURNACES

For Hard Coal

With Equalized Draft ❧ Patented August 5, 1890 ❧ Improved 1897

WITH STEEL RADIATOR

Cut shows Furnace without Casings, with Draw-Centre Grate and
Water Pan in Front

The upper edges of these drum-casings are securely adjusted to the flanges of an upper radiator, which, like the lower radiator, is cast in one piece. The illustration shows the three channels of communication between the inner and the outer drum; also the central check, which is cone-shaped on the upper side, and which equalizes the draft through the inner drum. The draft through the outer drum is equalized by means of a graduated Ring Check.

TOP VIEW OF LOWER RADIATOR

The inner drum casing forms a central smoke chamber, an outer smoke chamber being afforded by the space between the middle and the outer drum casing. These two smoke chambers *communicate freely with each other and with the fire pot*, and are *perfectly self-cleaning*. The passage of the draft through them is carefully regulated by self-cleaning checks, so proportioned as to obtain a perfectly *equalized draft* and a uniform distribution of heat over the entire radiating surface of the drums.

Between the central and the outer smoke chamber is an annular hot air space, to which the air to be heated passes freely through inclined passages formed in the lower radiator.

BOTTOM VIEW OF UPPER RADIATOR

The feed door neck is cast in one piece with the lower radiator, and also communicates with the outer smoke chamber, whereby any possibility of flame being blown out through the feed door is avoided. The Base Section of the PARAGON FURNACE is round in form. Cold air may be intro- duced at the sides, back, or from be- neath the furnace, at pleasure. The cold air openings in the base are provided with removable panels, which may be either open or close, as ordered. The ash pit door is so arranged that the draft may be regulated by either a ratchet or a chain.

FINISHED STEEL RADIATOR TOP VIEW. (SHOWING ANNULAR HOT-AIR SPACE.)

Points of Advantage

" PARAGON STEEL-PLATE FURNACES " possess three essential points of advantage:

FINISHED STEEL RADIATOR BOTTOM VIEW. (SHOWING FEED DOOR NECK AND DUST FLUE OPENING.)

1. The inner and outer Combustion Cham- bers communicate with each other in such a way that all the Radiating Surfaces of both Combustion Chambers are equally heated.

2. Both Combus- tion Chambers are absolutely self- cleaning.

3. Both the upper and lower Radiator Cast- , ings are made in one piece.

When any one offers you a three-drum Furnace said to be "as good as the PARAGON,"

PARAGON HOT AIR FURNACES

For Hard or Soft Coal

With Equalized Draft ❧ Improved January 2, 1895 ❧ Improved 1897

WITH CAST RADIATOR

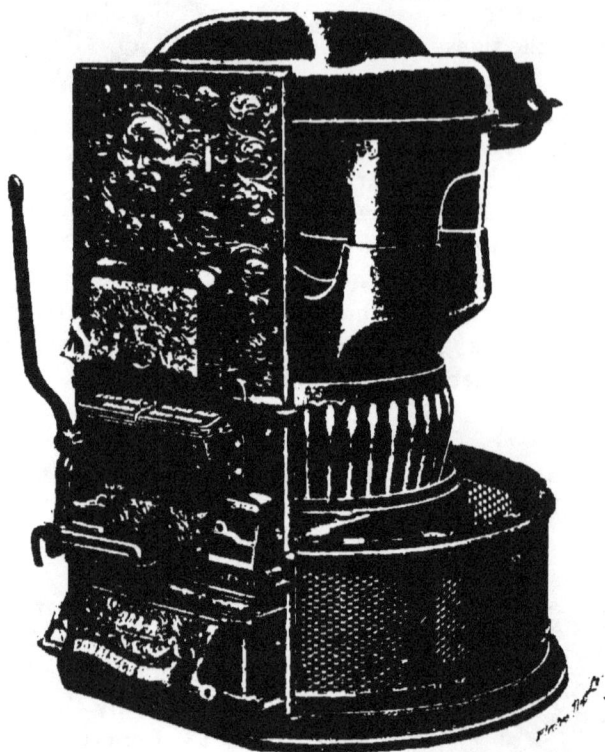

Cut shows Furnace without Casings, with Draw-Centre Grate and
Water Pan in Front

look for yourself, and see whether these points are met in its construction.   To experienced Furnace-setters, we do not need to say that there is *no other three-drum furnace made* that embraces the three points above named in its construction.   It has long been conceded that a perfect three-drum furnace must possess these requisites; but until the problem was solved in the PARAGON FURNACE, manufacturers thought that it was a practical impossibility to combine them.   The successful accomplishment of this feat has placed the PARAGON at the head of all furnaces of this class.

With Cast Radiator
SIX SIZES:

No. 428-A.   With High Front ; Diameter of Outer Casing, 28 inches; Diameter of Fire Pot, 16 inches.
No. 433-A.   With High Front ; Diameter of Outer Casing, 33 inches; Diameter of Fire Pot, 19 inches.
No. 436-A.   With High Front ; Diameter of Outer Casing, 36 inches; Diameter of Fire Pot, 21 inches.
No. 440-A.   With High Front ; Diameter of Outer Casing, 40 inches; Diameter of Fire Pot, 23 inches.
No. 444-A.   With High Front ; Diameter of Outer Casing, 44 inches; Diameter of Fire Pot, 25 inches.
No. 448-A.   With High Front ; Diameter of Outer Casing, 48 inches; Diameter of Fire Pot, 28 inches.

Points of Advantage

"PARAGON CAST RADIATOR FURNACES" embody the same advantages, in all essential particulars, that are possessed by the Steel-Plate Furnaces, especially in the equalization of the draft and the perfect utilization of fuel.   When used with hard coal, these Furnaces are perfectly self-cleaning.   When used with soft coal, they can be easily cleaned, when necessary, ample facilities being provided for the purpose.

Constructive Features of Cast Radiator Furnaces

The lower Radiator, which rests directly upon the Fire Pot, is cast in one piece with the Feed Door Neck. The part directly over fire is strongly corrugated, strengthening the radiator, as well as increasing its heating capacity. The inner smoke chamber is a cast iron cylinder that fits over the central opening in lower Radiator.

LOWER RADIATOR. TOP VIEW.
(SHOWING FEED DOOR NECK
AND CUP JOINTS.)

The outer smoke chamber is formed by a middle Radiator, a five-armed casting resting upon the lower Radiator. The two upright check-plates and the middle cone-check shown in the illustration perfectly equalize the draft in this form of the PARAGON FURNACE.

All smoke currents meet in the top Radiator smoke-channels, and pass together to the smoke outlet. All joints are deep cup joints, and are solidly packed with asbestos cement when the furnace is erected, making it absolutely gas-tight. The illustration shows the cast Radiator complete, a massive and substantial construction. While it was originally designed for use with soft coal, it is rapidly gaining favor among those who use hard coal or coke as fuel. The castings

MIDDLE RADIATOR SHOWN
IN POSITION OVER
LOWER RADIATOR

manifest astonishing durability. Some of the chief merits of the PARAGON FURNACES are enumerated in the pages that follow.

Great Radiating Power

THE PARAGON FURNACE possesses the *largest radiating surface* in proportion to grate surface, of any three-drum furnace yet made. The equalized draft renders every square inch of this surface *equally effective*. The result is three-fold: — an ample and constant supply of warm air; *equal wear* upon all parts of the furnace; and perfect utilization of the heat of the fuel. This means *superior economy*, *efficiency* and *durability*.

TOP RADIATOR
(BOTTOM VIEW)

No Heat Lost in Cellar

THE PARAGON FURNACE is *double cased throughout*, the air space between the casings serving as a non-conducting chamber, preventing loss of heat and increasing the efficiency of the furnace. What is desired is to heat the *house*, and *not the cellar*. This the PARAGON accomplishes.

CAST RADIATOR COMPLETE
(SHOWING FRONT CLEAN-OUT)

Effective Ventilation Secured

The ample provision made in the structure of the PARAGON for the admission of air, and for its rapid distribution in large volumes over the heating surfaces, ensures adequate ventilation. It brings about a constant influx of pure warm air into the rooms to be heated, which continually displaces an equal amount of vitiated air, and establishes the claim of the PARAGON to be regarded as an effective ventilating apparatus.

No Dry, Parched Air, but Pleasant Warmth

A large water pan is provided at the front of the furnace just where it can most easily be examined and most conveniently filled. Constant evaporation from the surface of the water contained therein furnishes the needed moisture. The water pan is protected from undue heat by interposing the dust flue between it and the fire pot; while the liability of the fire pot to burn out in front, where very little air comes into contact with it, is diminished by placing the water pan at that point.

Perfect Combustion

The construction of the PARAGON FURNACE, although so simple that it can be understood by a child, is nevertheless unsurpassed in securing perfect combustion. The draft is always *direct*, and perfectly *equalized* throughout. The fire can be kindled in about half the time required by other three-drum furnaces. The combustion will be found *perfectly equal* throughout the whole mass of fuel. This results in comparative freedom from clinker, and in thorough utilization of the fuel. A large amount of heat—a small residue of ashes and clinker—these are the results attained by the PARAGON FURNACE.

Ease of Management

The draft is regulated by raising or lowering the drop shutter in the ash pit door, and by closing or opening the Draft Check at the back of the furnace.

Both of these are held at any desired point by means of a ratchet; or if preferred, they may be connected with a chain, and thereby operated from the room above. In the PARAGON with Draw-Centre Grate, the whole surface of the grate is exposed to view upon opening the sliding fire doors.   By means of the poker, clinker can be easily removed from any part of the fire, and dropped through the centre of grate into the ash pit. The grate, in the larger sizes, is hung upon ball bearings and

PARAGON DRAFT CHECK

connected with a lever shaker. A person can operate this without stooping; and its action is so easy that a child can thoroughly shake the grate of the largest furnace.   The purchaser of the PARAGON has also the choice of an improved form of Triple X Grate, upon which a patent has been granted.   Both of these grates may be easily removed and replaced, as shown in the accompanying illustrations.

Paragon Draw-Centre Grate

With Ball Bearings

The illustration shows the ease with which the removal of this grate is accomplished. All clinker is driven to the centre of this grate. The connecting bar is lifted out of its socket in upright lever, and at once becomes a handle with which to pull out the centre-draw.   All sizes above 33 inches have ball bearings.

Paragon Triplex Grate
Patented October 1, 1895

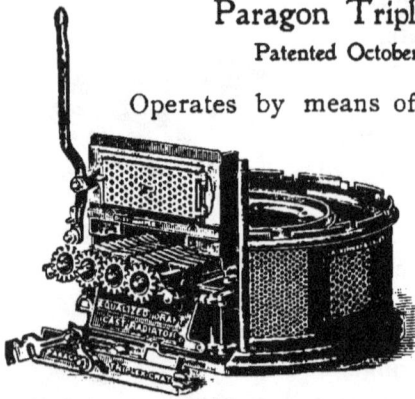

Operates by means of upright lever.   When the connecting bar is placed in the inner notch, the grate is shaken like an agitating grate.

When the bar is placed in the outer notch, the grate acts as a triplex grate, turning over at each movement of the lever.

Superior Cleanliness

The ash pit of the PARAGON is capacious, and the ash pit door both wide and deep, affording every facility for the easy removal of ashes.   A dust flue, placed *at the front* where it is needed (not at the back of the ash pit where it is liable soon to be choked up with refuse, and rendered useless), protects the operator from annoyance.   As has before been said, the drums are self-cleaning throughout, and require no attention.

Freedom from Gas

The PARAGON FURNACE is in this respect faultless. Its superiority in construction will be manifest to any one who will examine competing furnaces.   In the Paragon Furnace the lower radiator (sometimes called the "crab,") is *made in one piece*,—absolutely *jointless*.

In all other furnaces of this class, this "crab" is

usually made in *three parts*, never less than *two*. These parts, owing to their peculiar formation, expand and contract irregularly, becoming warped and distorted, and opening the joints between them at the very points at which gas and smoke are most liable to escape into the hot air chamber. No matter how carefully made and tightly cemented these joints at first may be, it is absolutely impossible to keep them tight in actual use. In such constructions, leakage of gas is *unavoidable*. In the Paragon lower radiator, it is *impossible*.

The upper radiator of the PARAGON is also made in one piece; and all joints are so formed that they can be packed gas tight with asbestos cement.

Adaptability

Either hard or soft coal or coke can be employed with satisfaction in the PARAGON FURNACE. No furnace now on the market can more successfully meet the varied requirements of different sections.

Superior Durability

Every part of the PARAGON FURNACE is skilfully proportioned in thickness to the amount of strain that it is required to endure. Not only is the Paragon heavier than other furnaces, but the extra weight of metal is placed *where it will do the most good*. The steel drums are *extra heavy*, with double riveted seams. The fire pot and lower radiator are *corrugated* throughout. This formation not only largely increases the radiating surface, but also reduces to a minimum the risk of cracking by fire.

In these, as in all other respects, a careful comparison of the PARAGON with all other furnaces is invited. The PARAGON is *no imitation of previous structures:* but in *merit* as in *originality*, it *leads them all.*

PARAGON COMBINATION FURNACES

Hot Water and Hot Air

WITH EITHER STEEL OR CAST RADIATOR

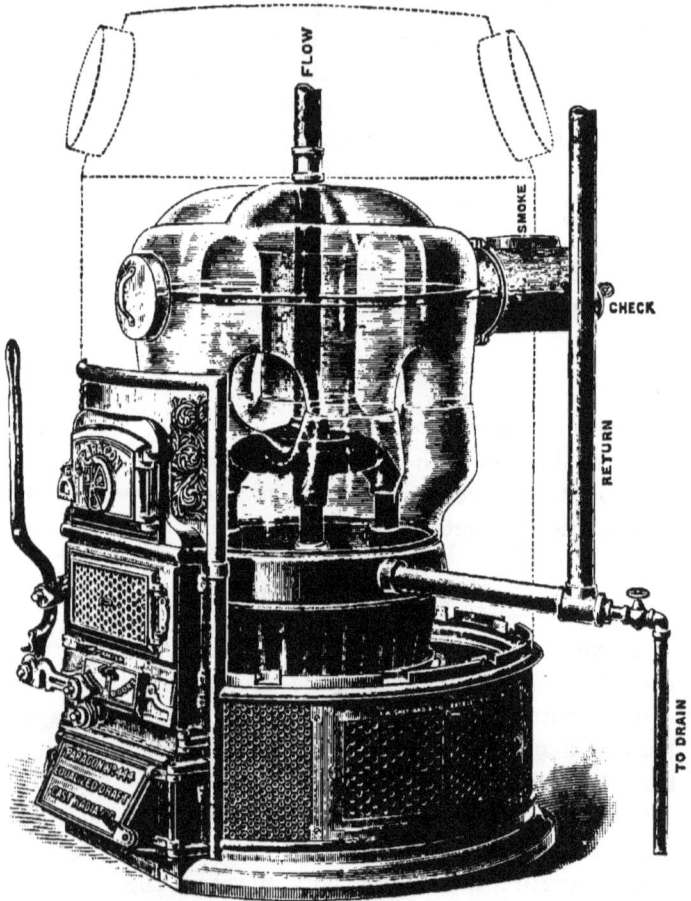

Paragon Combination, with Cast Radiator
Made also with Steel Radiator

Cut shows Improved Triplex Grate with Lever Shaker

Three Sizes:
With either Steel or Cast Radiator: 40 inch; 44 inch; 48 inch

Paragon Combination Furnace

A very efficient heating apparatus, enabling distant rooms to be thoroughly heated, that could not be reached from an ordinary hot-air furnace. This mode of heating also affords better ventilation through its direct warm-air supply than an ordinary hot-water apparatus.

Paragon Water Heater

The cut shows in detail the water heater used in all our Combination Furnaces. It is highly efficient. The return pipe may enter on either side; or two returns may be used.

Persons who are thinking of installing a Paragon Combination Furnace should write directly to us. We will furnish blanks, asking certain particulars, upon answer to which, we shall gladly give such information as will enable the work to be properly done by any suitably qualified mechanic. We make no charge for such service.

Paragon Brick-Set Furnaces
With Either Steel or Cast Radiator

Some of the advantages of the PARAGON BRICK-SET FURNACES are:

1. The front is both higher and wider than the furnace itself, and the latter can easily be removed through the front opening without disturbing the brickwork.

PARAGON BRICK-SET FURNACES

With Equalized Draft   Patented August 5, 1890

WITH EITHER STEEL OR CAST RADIATOR

Paragon Brick-Set Furnace ❧ Front View

With either Cast or Steel Radiator

Cut shows Furnace with Draw-Centre Grate

PARAGON BRICK-SET FURNACES

With Equalized Draft   Patented August 5, 1890

WITH EITHER STEEL OR CAST RADIATOR

Paragon Brick-Set, with Cast Radiator ❧ Side View

Made also with Wrought-Steel Radiator

Cut shows Furnace with Adjustable Cast-Iron Elbow

2. The front is in three parts, each of which is easily removed to inspect the interior brick chamber. No man-hole door is required with this furnace.

Paragon Hot-Air Furnaces
Estimated Heating Capacity Under Average Conditions
TABLE A

SIZE OF FURNACE	Diameter of Fire Pot	Heating Capacity in Cubic Feet when Divided into Rooms as in Residence	Heating Capacity in Cubic Feet Undivided as in Churches and Stores	Capacity Sufficient to Employ Hot-Air Pipes of Sizes below (OR EQUIVALENTS)	Cold-Air Duct to be Employed of Sizes below (OR EQUIVALENTS)	Size of Smoke Pipe Requir'd
Nos. 328 and 428 Paragon,	16 inches	12,000 to 16,000	15,000 to 18,000	4—9 inch	12 x 16 ins.	6 inch
Nos. 333 and 433 Paragon,	19 inches	15,000 to 20,000	20,000 to 25,000	4—10 inch	13 x 18 ins.	7 inch
Nos. 336 and 436	21 inches	18,000 to 25,000	25,000 to 30,000	5—10 inch	14 x 20 ins.	7 inch
Nos. 340, 440, 540, and 640	23 inches	25,000 to 35,000	35,000 to 45,000	4—12 inch	16 x 21 ins.	7 inch
Nos. 344, 444, 544, and 644	25 inches	30,000 to 40,000	45,000 to 60,000	6—12 inch	20 x 25 ins.	8 inch
Nos. 348, 448, 548, and 648	28 inches	40,000 to 60,000	60,000 to 80,000	5—14 inch	22 x 26 ins.	8 inch

COMBINATION PARAGON FURNACES	Heating Capacity in Sq. Ft. of Direct Radiation	Heating Capacity in Cubic Feet by Hot Water when Divided	Heating Capacity in Cubic Feet by Hot Air when Divided	Diameter of Flow Pipe	Diameter of Return Pipe	Size of Smoke Pipe
40 inch sizes............	400	12,000 to 14,000	10,000 to 12,000	2½ inches	2½ inches	7 inch
44-inch sizes............	450	14,000 to 16,000	12,000 to 15,000	2½ inches	2½ inches	8 inch
48 inch sizes............	525	16,000 to 22,000	15,000 to 18,000	3 inches	3 inches	8 inch

NOTE.—While under average conditions, if the furnaces are properly set and judiciously managed, these estimates will be found to approximate closely to correctness, yet the exposures of buildings vary so greatly that another table, designated as Table C, which will be found on page 34 has been prepared, whereby the size of furnace that is best suited to any given case may be more accurately determined.

3. Grates are easily removed; and all grates, whether draw-centre or triplex, are operated by a lever shaker.

4. The interior construction is the same as in the portable forms of the PARAGON FURNACE. Nothing better has ever been devised.

5. If the water pan in front is not desired, a fretwork door will be fitted in its place, as shown in cut on page 30, and a different water pan substituted for it, which can be walled up in any part of the brickwork where the dealer prefers to place it.

Unless otherwise specified, these furnaces will always be shipped with water pan in front, and with ball-bearing draw-centre grate. Made in three sizes, Nos. 540, 544 and 548, with steel radiator; and Nos. 640 644 and 648, with cast radiator. Also with combination hot-water attachment for any size named above.

Particular attention is called to the fact that to get the best results from any furnace, the cold-air supply should be taken from outside of the building; and that the cold air duct should be of full size indicated in the above tables until it reaches the furnace, where if it does not enter a pit beneath the furnace, it should be divided into two branches, each the full size of the cold air opening in the base of the furnace, and so carried around to each side of the furnace.

Instructions for Use of " Table C "

To find size of furnace or furnaces best adapted to any building, ascertain the contents of building in cubic feet, the number of square feet of exposed wall surface, and the number of square feet of glass in windows. Then calculate by the following rule:

RULE—1. Multiply the cubic feet of contents by

Index of Heating Capacities. Table C

Copyright 1892, 1895 and 1897

(FOR INSTRUCTIONS SEE PAGES 33 AND 34)

Paragon Hot-Air Furnaces

SIZES (ANY STYLE)

28 inch	33 inch	36 inch	40 inch	44 inch	48 inch
32	37	45	53	61	74
33	38	46	54	62	75
34	39	46	55	63	76
35	40	47	56	64	77
36	41	49	57	65	78
37	42	50	58	66	79
38	43	51	59	67	80
39	44	52	60	68	81
40	45	53	61	69	82
41	46	54	62	70	83
42	47	55	63	71	84
	48	56	64	72	85
	49		65	73	86
				74	87
				75	88
				76	
				77	

$\frac{8}{10}$, the square feet of exposed wall by 4, and the square feet of glass by 40, and add together the several products.

2. Divide the sum by 600, if the space in building is divided into rooms as in a residence; or by 800, if the space is undivided, as in a church or store.

3. Look in Table C for the quotient thus arrived at. The furnace indicated in the column in which such number is found, is the furnace to be recommended for the building in question.

NOTE.—If the quotient number be found in *more than one column*, it indicates that either furnace indicated will do the work. The larger of the two is, however, to be preferred, as being the more durable.

Note.—If the quotient number be *higher* than the highest number in Table C, *two* or *more* furnaces will be required, of such sizes as are indicated by the *highest component numbers* of the quotient number that appear in the table.

> EXAMPLES.—1. A residence has 13,000 cubic feet of space, 840 square feet of exposed wall, and 230 square feet of glass. Find size of suitable furnace.

$$13,000 \times \tfrac{8}{10} = 10,400$$
$$840 \times 4 = 3,360$$
$$230 \times 40 = 9,200$$
$$22,960 \div 600 = 38 +$$
Quotient.

In Table C the quotient number 38 is found both in the column headed "28-inch size," and in that headed "33-inch size." This indicates that while the 28-inch size will do the work, the 33-inch size is preferable.

2. A residence has 31,200 cubic feet of space, 1,425 square feet exposed wall, and 340 square feet glass. Find suitable furnace or furnaces.

$$31,200 \times \tfrac{8}{10} = 24.960$$
$$1,425 \times 4 = 5,700$$
$$340 \times 40 = 13,600$$
$$44,260 \div 600 = 73 +$$
Quotient.

The quotient number appears in column headed "44-inch size." The component numbers 33 and 40, the sum of which equals 73, appear in columns headed respectively "28-inch" and "33-inch" sizes. Therefore, you may use either one 44-inch PARAGON, or if two furnaces can more conveniently be employed, one 28-inch and one 33-inch PARAGON instead.

3. A church has 143,000 cubic feet space, 5,600 square feet exposed wall, and 1,150 square feet glass. Find suitable furnaces.

$$143,000 \quad \tfrac{8}{10} = 114,400$$
$$5,600 \times 4 = 22,400$$
$$1,150 \times 40 = 46,000$$

$$\overline{}$$
$$182,800 \div 800 = 228 +$$
Quotient.

The quotient is higher than appears in table. The component number 57 appears in column headed "40-inch size." (57 × 4 = 228.) So that 4—40-inch PARAGONS will do the work. The component number 76 also appears in columns headed "44 inch" and "48-inch" sizes, (76 × 3 = 228). Hence three 44-inch, or preferably three 48-inch PARAGONS may be employed.

NOTE

The estimates of heating capacity of PARAGON FURNACES herein given, are all calculated upon the basis of the winter temperature of the city of Philadelphia, where the extreme low temperature never exceeds 5° below zero. They may be used for any locality in which the thermometer is not known to fall below 7° below zero; such cities, for example, as Boston, New York, Baltimore, Washington, D. C., Cincinnati, Louisville and Atlanta. For localities in Western Maryland, West Virginia, Southwestern Virginia, Eastern Tennessee and Western North Carolina, the estimates of Table C also hold good.

For localities in which temperatures ranging from 7° to 20° below zero are occasionally experienced, 20 per cent. additional capacity should be provided; as, for example, Pittsburg, Pa., Portland, Me., Burlington, Vt., Albany, Buffalo, Chicago, Detroit, Indianapolis and St. Louis.

For localities in which temperatures as low as from 25° to 38° below zero sometimes occur, as, for example, St. Paul, Minneapolis, Duluth and Milwaukee, add 40 per cent. to the figures given in the table.

On the other hand, where the winter temperature never falls more than from 6° to 16° below the freezing point, or say to 16° above zero, one-third less heating capacity will be necessary, and from 30 to 35 per cent. may be deducted from the figures given in the table.

Such cities as Wilmington, N. C., Charleston, S. C., and Savannah, Ga., fall into this class.

Directions for Using Paragon Furnaces

Under the heading "Furnace Management," will be found some general comments that will be of service to any one who has charge of a hot air furnace. Some specific directions, applicable to the "PARAGON FURNACE," may also be useful.

To Kindle the Fire

Have the pipe and chimney unobstructed, the grate in its proper position, and the ash pit free from ashes and refuse. Cover the grate well with shavings and small chips. Have a good supply of larger wood ready. Before lighting fire, close check draft at back of furnace, and lift the shutter in ash pit door as far as the ratchet will permit. As soon as the shavings and chips are well kindled, put in the larger wood, a few pieces at a time, until there is a good fire. Put on not more than three or four shovelsful of coal at first. When this is fully ignited, add as many more. After this second supply has been thoroughly kindled, fill up the fire pot, and close the shutter in ash pit door. The check draft at the back must not be opened until all the gas from the fresh coal has passed off. Then, if desired, the fire may be checked enough to keep it burning moderately, although steadily.

To Care for the Furnace

Remove the ashes once every twenty-four hours, as the air must circulate freely under the grate, or it will burn out.

Do not attempt to clear a *low fire*. Let it first burn up for fifteen or twenty minutes. When shaking the grate close the damper or ash pit door and open the dust damper; this will keep the dust from coming out into the cellar.

Let the register-wheel in the feed door remain open except when starting the fire; this admits air over the surface of the fire and will cause the coal gas to ignite and burn.

When putting coal into the furnace close the check damper in rear of furnace and the damper on ash pit door, otherwise gas and smoke may flow out of the feed door and get into the rooms above.

Never open the feed door except when putting in coal. To check the fire, open the damper in rear of furnace and close the one on ash pit door.

Keep the fire pot filled with coal, even with the feed door, and in cold weather heap it up; there is no economy in running a small fire.

To keep the fire over night, fill the fire pot rounding full, and open the check damper in rear to the second notch, and the ash pit damper to the first notch.

The cold air pipe damper must never be entirely closed; the supply of cold air should be governed by the temperature of air coming through the register. If cold air comes up any of the registers while there is a good fire, reduce the air supply; if any register does not emit any air, increase the supply. The cold air supply must be governed by the weather.

Keep the water pan full of water. If the pan is allowed to get dry it should be taken out and washed clean, otherwise it may give an unpleasant odor in the house which is often mistaken for coal gas.

To remove the clinkers from the draw centre grate, lift the hook from the shaker lever and pull the centre · of the grate toward you, and then put the poker through the clinker doors and knock the dead ashes and clinkers through the centre of the grate to the ash pit; then shove the centre of the grate to its place and shake same.

For furnaces with the triplex grate the foregoing directions apply, except as to the use of the grate. Large coal must never be used with the triplex grate. To *agitate* the grate, put the connecting lever upon the pin at the bottom of the upright lever at the *inner notch*. When the clinker-clearing or dumping movement, characteristic of the triplex grate, is desired, move the lever to the *outer notch*. These movements will always serve to keep a clear and bright fire with hard coal of good quality and proper size. With too large coal, or coal of poor quality, clinker is apt to form in the centre of the fire, which is less easily removed by the triplex than by the draw centre grate.

A furnace should be examined every spring, by a competent furnace man, who should clean the smoke pipe and see that everything is in good working order. When soft coal is used it may be found desirable to have the smoke-pipe taken down and cleaned more than once during the winter. If soft coal is to be the fuel used, it should be stated when the order is given, that the drum checks may be adjusted so as to prevent any clogging of the furnace by soot.

Paragon Combination Furnaces

The cuts on pages 43–45 represent the floor plans of a city house heated with a No. 340 PARAGON FURNACE with water heating attachment. The house being narrow with three stories and basement, having occupied rooms requiring heat on each floor, with main stairway between the front and rear parts, the illustrations show the facility with which distant rooms can be heated from one furnace with water heater in cases in which two furnaces would be objectionable. The cuts show how the front building is heated by hot air, a large register being opened in the back part of hall near the staircase, to preserve the proper balance between the hot air and the hot water parts of the system. The main flow pipe and part of the return pipes are located near the basement ceiling. The return pipe drops to the floor at the radiator in the dining room and runs above the floor through the stairway and cellar to a point opposite the furnace, where it passes into a brick trench below the cellar floor (shown by dotted line) to the furnace, and up into the water heater. The draw-off cock is placed at the partition between cellar and stairway, the waste from same running into a gutter in the area at side of house. Two cold air ducts are shown, opening at opposite sides of house. The one on the side against which wind is blowing at any particular time is opened, and the other is then kept closed.

Directions for Setting Paragon Portable Furnaces

If furnace is to stand over a pit, carry up a central pier of brick work under ash pit, to support weight of furnace. If it is to stand on cellar floor, place under it a course of brick, carefully leveled and cemented on top, to prevent dust arising from floor of cellar. Have the furnace base *perfectly level*, as otherwise the furnace sections, when erected, will not stand plumb, and may not fit well. See that the ash pit ring, in the 44-inch and 48-inch sizes is put in place *with the lug in front*. On the 40-inch and smaller sizes this ring is bolted fast. Cement this ring thoroughly, and then place the fire pot on ash pit ring, so that notches in fire pot cover projections on ring. Fill in around bottom of fire pot with asbestos cement, a can of which accompanies each furnace. Put dust damper in place, and put the lower dust pipe on the oval collar. Then lift the drum section up on fire pot. Secure the upper dust flue pipe with bolts to oval collar under neck of furnace, and set the drum section in place, with the notches in lower radiator covering the lugs on the fire pot, slipping the upper dust pipe over the lower one. Fill up joint between fire pot and upper radiator with asbestos cement.

Then put on lower galvanized casing, and draw it up neatly to place, bolting it fast to front. Put on the lower inside casing and next fix the lower casing ring (with two flanges on top) in position. Next put on the upper inside casing, taking care to see that the hole in casing around smoke collar is large enough to allow for free expansion. Then put the upper galvanized casing neatly into place, bolting it on one side to the upper front and then pulling it up to place on the other side by means of wire passed through the bolt holes of the casing. Put on the upper casing ring. Bolt on the draft check, and also the upper front. Have all joints properly cemented. Put on the dome top with outlets, and the furnace will be ready to connect with the hot air pipes.

Paragon Brick Set Furnaces

Any person accustomed to setting furnaces in brick work will find no special instructions needed. The furnace should be firmly bedded in cement, so that the projections on the bottom may hold the furnace firmly in position when the grate is being shaken. The front opening should be covered by a bar, and the size so arranged as to make a neat finish behind the front moulding. The front is not fastened to the brick work, and may easily be removed at any time.

Measurements of Casings, etc., for Paragon Hot Air Furnaces

Measures of Circumference are neat, and do not include lap for seaming or riveting. The Swedging of Galvanized Casings should stop 2 inches from each end, to finish against Furnace Front.		Lower Galvanized Casing	Upper Galvanized Casing	Dome Rims	Lower Inside Casing	Upper Inside Casing	Height of Furnace Uncased	Height of Furnace Finished	Fire Pot Diameter	Depth of Fire	Size of Smoke Pipe
Nos. 328 and 428	Gauge of Sheet Iron..	27	27	27	27	27	4 ft. 6 in.	5 ft. 5 in.	16 in.	14½ in.	6 in.
	Height.............	21 in.	16½ in.	2¾ in.	21 in.	16⅜ in.					
	Circumference..	6 ft. 3½ in.	6 ft. 3½ in.	7 ft. 3¾ in.	5 ft. 8 in.	6 ft. 9¼ in.					
Nos. 333 and 433	Gauge of Sheet Iron..	27	27	27	27	27	4 ft. 8 in.	5 ft. 8 in.	19 in.	15 in.	7 in.
	Height.............	24 in.	14⅞ in.	2¾ in.	24 in.	14⅞ in.					
	Circumference........	7 ft. 6⅜ in.	7 ft. 6⅜ in.	8 ft. 7⅝ in.	6 ft. 11 in.	8 ft 0⅝ in.					
Nos. 336 and 436	Gauge of Sheet Iron..	27	27	27	27	27	4 ft. 10 in.	5 ft. 8½ in.	21 in.	15 in.	7 in.
	Height.............	24 in.	14⅞ in.	2¾ in.	24 in.	14⅞ in.					
	Circumference........	8 ft.	8 ft	9 ft. 5 in.	7 ft. 8½ in.	8 ft. 10⅜ in					
Nos. 340 and 440	Gauge of Sheet Iron..	24	24	24	24	24	5 ft.	5 ft. 9 in.	23 in.	15½ in	7 in.
	Height.............	24 in.	14¾ in.	2¾ in.	24 in.	14¾ in.					
	Circumference........	8 ft. 9⅜ in.	8 ft. 9⅜ in.	10 ft. 5½ in.	8 ft. 6 in.	9 ft. 11 in.					
Nos. 344 and 444	Gauge of Sheet Iron..	24	24	24	24	24	5 ft. 0¼ in	5 ft. 11½ in.	25 in.	15½ in.	8 in.
	Height.............	24 in.	15⅜ in.	2¾ in.	24 in.	15⅜ in.					
	Circumference........	10 ft.	10 ft.	11 ft. 5½ in.	9 ft. 5¾ in.	10 ft. 10¾ in					
Nos. 348 and 448	Gauge of Sheet Iron..	24	24	24	24	24	5 ft. 2 in.	6 ft. 5 in.	28 in.	16 in.	8 in.
	Height.....	24 in.	15⅞ in.	2¾ in.	24 in.	15⅞ in.					
	Circumference........	10 ft. 7½ in.	10 ft. 7½ in.	12 ft. 7½ in.	10 ft.	12 ft. 0½ in.					

SMALLEST BRICK CHAMBER IN WHICH PARAGON FURNACES CAN BE SET: INSIDE MEASUREMENTS: ADJUSTABLE CAST ELBOW NOT INCLUDED.

Nos. 540 and 640; 2 ft. 9½ in. wide; 3 ft. 6 in. deep; 5 ft. 2 in. high.
Nos. 544 and 644; 3 ft. 1 in. wide; 3 ft. 10 in. deep; 5 ft. 3 in. high.
Nos. 548 and 648; 3 ft. 4½ in. wide; 4 ft. 2 in. deep, 5 ft. 7½ in. high.

House Heated by Paragon Combination Furnace

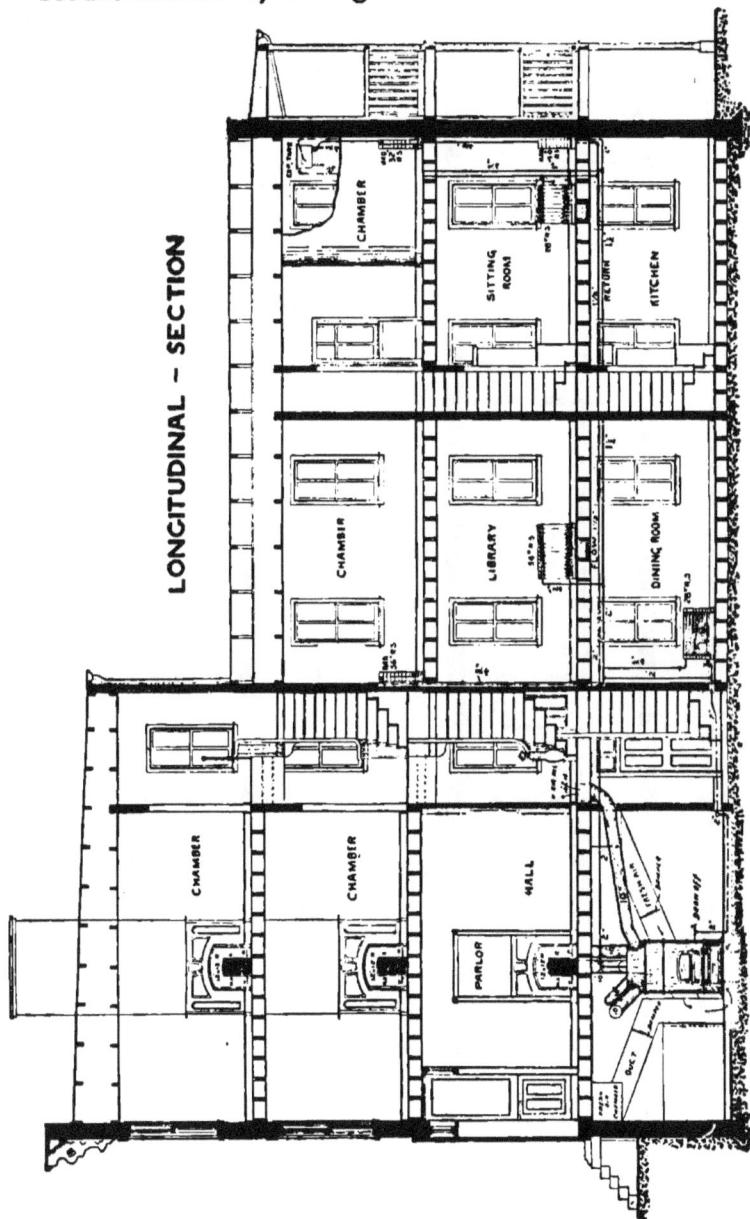

Longitudinal Section

House Heated by Paragon Combination Furnace

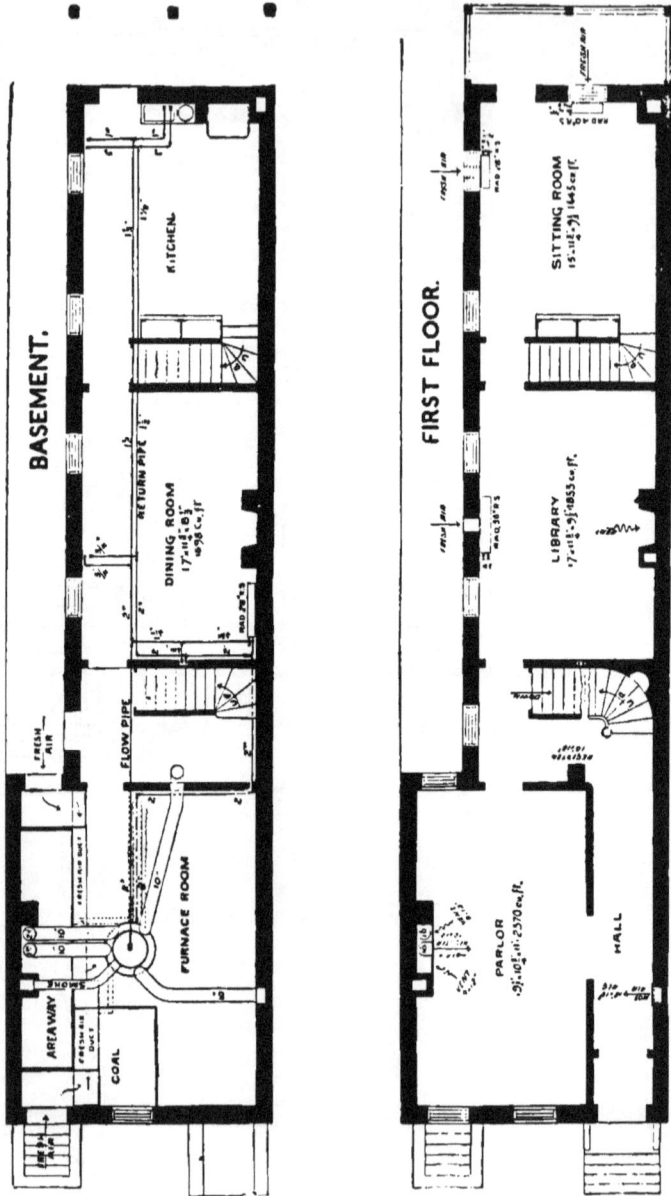

Basement and Floor Plans

House Heated by Paragon Combination Furnace

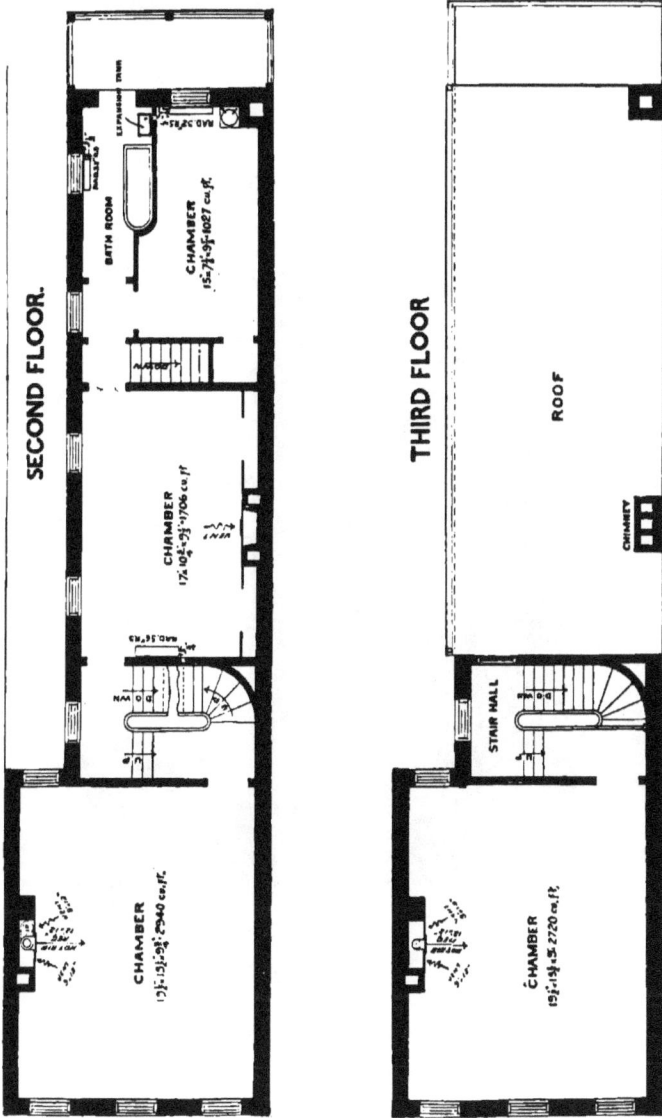

SECOND FLOOR.

THIRD FLOOR

BATH ROOM

CHAMBER
15'-7½'-9'-1027 cu.ft.

CHAMBER
17'-10½'-9½'-1706 cu.ft.

CHAMBER
13'-13'-9½' 2940 cu.ft.

ROOF

CHIMNEY

STAIR HALL

CHAMBER
19'-19'-12 2720 cu.ft.

RAD. 36" H.

RAD. 36" H.

EXPANSION TANK

DOWN

Second and Third Floor Plans

PARAGON HOT AIR FURNACE

With or without Water Heating Attachment

Cut shows Finished Furnace with Water Pan in Front